ENERGY
UNLIMITED

4 STEPS TO 100%
RENEWABLE ENERGY

BARBARA ALBERT

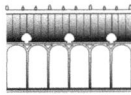

LONGUEVILLE
MEDIA

First published 2017 for Barbara Albert by

LONGUEVILLE
MEDIA

Longueville Media Pty Ltd
PO Box 205
Haberfield NSW 2045 Australia
www.longmedia.com.au
info@longmedia.com.au
Tel. +61 2 9362 8441

For the National Library of Australia Cataloguing-in-Publication entry see nla.gov.au

978-0-9945777-0-2 (print)

978-0-9945777-1-9 (eBook)

My Commitment to a More Sustainable Future

Ten per cent of author royalties will be donated to FONA, a for-purpose organisation, fostering new approaches to international development. It builds upon the natural strengths of communities, connecting them with global expertise to find innovative and sustainable solutions.

The money goes directly to FONA's Education Centre in Nepal. The project aims to create a new blueprint for educational and community infrastructure for remote communities, applying a whole-systems approach. The Centre was designed using natural, renewable solutions with solar for energy, water harvesting, and a biogas sanitation system. For more details about the project visit www.fona.org.au/programs/puranojhangajolieducationcentre

Lismore City Council (LCC) boldly committed to become self-sufficient in electricity from renewable sources by 2023. Initially the task seemed daunting, but with Barbara's help it all seemed straightforward and achievable. Our key stakeholders were involved and brought on board throughout the process with interactive workshops, one-on-one meetings, and site visits. What I liked was that Barbara and her team delivered a strategy and action plan that was very clear and, most of all, practical. This has enabled council to move straight into implementing actions. I would highly recommend her for any project where the client seeks sustainability leadership through their carbon or energy management.

– **Sharyn Hunnisett, Environmental Strategies Officer, Lismore City Council**

Lismore City Council has set itself the ambitious goal to be self-sufficient in electricity use by 2023. Barbara and her team were selected to analyse our energy situation, consult with staff and councillors, and deliver our Renewable Energy Master Plan. She was professional and engaging in her delivery and provided great value to council. I would recommend her to any organisation that is looking for a long-term renewable energy plan.

– **Gary Murphy, General Manager, Lismore City Council**

In 2015, Coffs Harbour City Council engaged and worked with Barbara in the preparation of the Coffs Harbour Emissions Reduction Plan. It was interesting to see all the options council staff had in terms of energy efficiency and renewable energy, and to watch the progress on how unsuitable options were filtered out to arrive at the business cases that made the most sense from an environmental and financial perspective. Barbara and her team made the target and interim targets tangible, as well as aligning the targets with council's existing strategy and plans. It was a pleasure to work with Barbara and her team, and I would recommend her to other organisations moving down this same path.

– **Chris Chapman, Director Sustainable Communities, Coffs Harbour City Council**

We at RFBI have dealt with Barbara over recent years and have found her most efficient and professional. Her consulting and advice are always of the highest professional standards. Barbara is an expert when it comes to helping organisations develop and implement their sustainability strategy and she is particularly knowledgeable in the fields of energy and carbon management. RFBI staff have also attended energy management courses run by Barbara and found the information and follow-up support invaluable.

– Steve Ellitt, Assets & Contracts Manager, RFBI

Barbara is a professional, engaging renewable energy expert, the right sort of person to work with if you want to achieve transformational change in pragmatic steps. She spends time understanding the organisation's current state and its readiness for change before identifying options for the future and steps to get there. Both a supportive and challenging mentor, Barbara is the ideal partner for organisations keen to move to 100% renewables in a way that best suits them.

– Julia Seddon, Head of Business Sustainability, Inghams Enterprises

At times during our sustainability journey we have needed leadership and guidance to achieve our goals. To this end we are very proud to be Australia's first and only NCOS-certified Carbon Neutral Registered Club. We could not have achieved this without the expertise, leadership, and guidance we received from Barbara Albert over the past years. Barbara has mentored us to become sustainability leaders in our industry and our community. Barbara provided a seamless and turnkey service for our club to achieve our Carbon Neutral goals. I would have no hesitation in recommending Barbara as your sustainability partner.

– Matt O'Hara, Chief Executive Officer, Oak Flats Bowling & Recreation Club and Illawarra Yacht Club

To my husband, Christoph Strizik

Contents

I'd put my money on the sun and solar energy. What a source of power! I hope we don't have to wait until oil and coal run out before we tackle that.

Thomas Edison, 1931

Why fossil fuels are bad

Burning fossil fuels releases the carbon that has been locked away in the Earth's crust for millions of years into the atmosphere, where it traps the sun's heat and contributes to climate change. Weather extremes already affect millions of people. Climate change puts food, water, and agricultural supply chains at risk, is responsible for the mass extinction of many species, and causes many diseases to spread.

It raises sea levels, potentially displacing millions of people. It increases the occurrence of heatwaves, droughts, and changes in rainfall patterns. And with more and more fossil fuels being burnt, the threats to our agriculture, vulnerable countries, and our entire society will get even worse.

According to the World Economic Forum's Global Risk Report,[1] extreme weather events, food crises, energy price shocks, and man-made environmental catastrophes like oil spills rank amongst the top worldwide risks. The failure of climate change mitigation and adaptation was listed as the most impactful risk for the years to come, ahead of weapons of mass destruction (ranked second) and water crises (ranked third).

In addition to contributing to climate change, our current energy production and use releases millions of tonnes of pollutants each year. Poor air quality is the world's fourth-largest threat to human health, behind high blood pressure, dietary risks, and smoking. According to the World Health Organization (WHO), 6.5 million deaths each year can be attributed to air pollution, much greater than the number from HIV/AIDS, tuberculosis, and road injuries combined.[2]

Foreword

Since 2008, the cost of solar PV panels has fallen by 80%. During the same period the cost of energy has continued to rise. As a result there has been a big surge in the installation of solar PV panels.

With the introduction of battery storage this number is set to rise even further. For the first time many companies will soon be able to capture unused solar energy at a truly economical cost.

These changes in the clean energy sector are happening far quicker than many people predicted. It's become obvious that the transition to cleaner power will be good for the business bottom line as well as our economy.

As the host of *Smart Money* on Sky News Business Channel, I've visited many companies that are slashing power bills with solar power and energy efficiency. But for every business that's saving money in this way, there are many others yet to make the move.

This is where Barbara Albert's book can be of real assistance. It provides a pragmatic road map for companies and organisations who want to transition to clean energy. It's also a useful tool for companies who want to tackle climate change by reducing their greenhouse emissions.

From reading this book, it's clear that Barbara Albert has experience with renewable energy. As she rightly points out, every organisation will have to wean itself off fossil fuels and convert to renewable energy. It's not a question of if, but when.

If your organisation wants to be part of this transition in a cost effective way, this book is a good place to start. Packed with information, case studies and graphics, it's a useful resource for managers who want to make their organisation more energy savvy and more sustainable.

Many business owners have found that utilising renewable energy is a great way to future-proof their businesses against rising energy prices. If your organisation wants to do the same, you'd be well advised to read this book.

Jon Dee
Anchor Host of Smart Money, Sky News Business Channel
Founder and Managing Director, DoSomething

Introduction

In 1882, the Edison Illuminating Company, headed by Thomas Edison, built the first central coal-fired power plant just south of Fulton Street in Manhattan. Being an inventor and an astute businessman, Edison created the plant to provide electricity to users of the incandescent light bulbs he had previously developed.

The day the power plant went live, Edison and his engineer synchronised their watches. Then, Edison and his board of directors went to the Wall Street office of JP Morgan. At exactly 3:00 pm, the engineer closed the plant's circuit breaker and Edison turned on the lights.[3] The age of electricity was born.

Having access to electricity brought unprecedented affluence to a huge number of people. In fact, much of our economic and social progress in the twentieth century was achieved through the energy contained in fossil fuels like coal, oil, and gas.

The power plant that Edison created was a prototype for all coal-fired power plants to follow. However, despite being the godfather of electricity-intensive living, Edison was also a green pioneer whose ideas still resonate today.

In 1901, he patented the first alkaline battery and hoped that electric cars would become the standard.[4] He had great plans to power rural homes with a combination of windmills, small electric generators, and battery storage.

Edison also began to understand that fossil fuels would not last forever and told his friends Henry Ford and Harvey Firestone, in 1931, shortly before he died:

'We are like tenant farmers chopping down the fence around our house for fuel when we should be using Nature's inexhaustible sources of energy – sun, wind and tide ... I'd put my money on the sun and solar energy. What a source of power! I hope we don't have to wait until oil and coal run out before we tackle that.'[5]

Unfortunately, by that time, people were accustomed to burning fossil fuels to obtain energy. Most countries continued to use fossil fuels for their energy needs into the early twenty-first century, building new coal-fired power plants and relying on importing oil for their transport needs, even though it became increasingly clear that burning fossil fuels was a major contributor to polluting and warming up our planet. At least 60% of our worldwide greenhouse gas emissions come from our energy use.[6]

A silent revolution

Things continued like this for decades but, just as we replaced typewriters with computers, horse-drawn carriages with cars, and landlines with smartphones, we are now transitioning to a new energy model that decarbonises, decentralises, and digitalises our energy supply and demand.

The new energy architecture is moving away from fossil fuels and centralised power generation to energy produced from renewable sources at or near where it is consumed. End users can now participate and invest in renewable power generation both directly and through innovative financing options. They can generate, store, manage, and trade energy.

Changes in the energy space and decreases in the cost of renewables are happening much faster than predicted, and investor preference has shifted from the old energy system to a clean one.

By 2016, renewables had outpaced fossil fuels for net investment in global power capacity additions for the sixth consecutive year.[7] There was significant growth in the development of new renewable energy in all regions, with China being the largest investor, followed by the USA, Japan, the United Kingdom, India, and Germany.[8]

In Australia, more than 15% of Australian households have installed solar PV on their roofs, double the residential solar PV penetration rates of the next country (Belgium), and more than three times the penetration of Germany and the U.K.[9] These are average figures – a few postcodes in Australia are already at 80%, which is indicative of the future.

In many parts of the world, people are switching to less carbon-intensive modes of transport, and they are also increasingly changing their fossil fuel-powered cars to electric ones. In April 2016, nearly 400,000 people placed orders for the new Tesla Model 3 electric vehicle that had not even been produced yet.[10]

In Norway, more than 30% of new cars purchased are electric and run using renewable hydroelectricity. Hydrogen vehicles are also emerging, with major manufacturers making production models. Autonomous vehicles are also flagged to radically change the way we consider and use vehicles.

In many European cities, rather than using cars, people are electing to walk, cycle, and use public transport options. One prominent example is Amsterdam which, since the 1970s, has changed from a car-dominated to a bicycle-dominated city.

In Germany, many people now rent cars when they need one, instead of following the traditional ownership model. The proliferation of shared-economy models, like Uber, is also evident. In Vancouver, bicycle traffic will continue to grow quickly over the coming decades as the city is aiming to increase the share of sustainable transportation modes to 50% by 2040.[11]

So, what has changed, starting in the early years of the twenty-first century to now?

A new, sustainable energy future

In May 2015, Pope Francis issued a papal letter 'On the care for our common home'.[12] In the letter, he warns about the risks of climate change and urges protection of the Earth, stating that, 'fossil fuels have to be substituted and sources of renewable energy developed'.

In September 2015, the United Nations adopted 17 Sustainable Development Goals (SDGs)[13] to end poverty, protect the planet, and ensure prosperity for all, as part of a new sustainable development agenda. The SDGs were largely influenced by the quote of UN Secretary-

General Ban Ki-moon that 'there can be no Plan B, because there is no Planet B'.

Each goal has specific targets to be achieved by 2030. Goal number seven, 'Affordable and clean energy', and goal number 13, 'Climate action', directly relate to the deployment of renewables. For us to reach these goals, the private sector needs to do its part, and many organisations worldwide have started to structure their sustainability initiatives around those 17 goals.

In December 2015, representatives from 195 nations adopted the first-ever universal, legally binding global climate deal, in Paris, at the 21st Conference of the Parties (COP21).[14] The historic agreement charted a new course in the global climate effort to accelerate the transition to a sustainable, low-carbon future.

The agreement's central aim is to keep the global temperature increase below 2 degrees Celsius, and to pursue efforts to limit it even further, to 1.5 degrees Celsius.[15] To meet this target, the scientific consensus is that we must achieve net zero emissions, or 'emissions neutrality', by 2050.

The Paris climate agreement has wide-ranging implications. In the wake of COP21, the International Energy Agency slashed its forecast of coal demand growth[16] and Saudi Arabia began preparing itself for a non-oil economy.[17] The agreement also recognises the role of cities and the private sector to scale up their efforts and support actions to reduce emissions.

Renewable energy technology has become mature and affordable, and, in some instances, does not even require upfront capital. It has widespread political support, and more and more organisations are embracing it.

Solar and wind energy prices have dropped like a stone and, in most regions, are cheaper than fossil fuels – solar energy alone has dipped more than 50% in price since 2008.[18] It is likely that batteries will experience similar price falls.

Deploying renewables is good for economic growth and our jobs. In January 2016, the International Renewable Energy Agency (IRENA) released a paper titled 'Renewable Energy Benefits: Measuring the Economics'.[19] According to the report, if the world doubled its current market share of renewable energy by 2030, global GDP could increase by 1.1%, which is equivalent to US$1.3 trillion in growth. It would also increase global welfare by 2.7% and create 24.4 million direct and indirect jobs.

Whole countries have recognised these new trends and are embarking on a path to more renewables. Germany is currently demonstrating how a country can transition away from fossil fuels to renewables whilst maintaining its thriving industrial economy through its *Energiewende* program, the largest post-war infrastructure project.

Costa Rica was powered by 99% renewable energy in 2015 and has pledged to become fully renewable by 2021. Uruguay was powered by 95% renewables in 2015, and Nicaragua by 54%, with plans to be 90% renewable by 2020.[20] Sweden plans on being 100% renewable by 2040, and the state of Hawaii has passed a mandate to reach 100% renewable electricity by 2045.

Iceland, with its huge hydroelectric and geothermal resources, already has 100% renewable electricity and 87% renewable heat. Denmark, with no hydro, is on track to achieve its target of 100% renewable electricity and heat by 2035.[21] It has an overarching goal of being entirely fossil fuel-free by 2050.

On a city level, there is an even bigger revolution underway. In Australia, municipalities like Adelaide aim to achieve net zero emissions by 2050, Lismore City Council wants to generate all its electricity needs from renewable sources by 2023, and Coffs Harbour City Council want to be fully renewable by 2030. The Australian Capital Territory is well on its way to achieving 100% renewable energy by 2020, and many more are currently evaluating new and ambitious renewable energy targets.

In Germany, more than 150 municipalities and regions are working to deliver a 100% renewable energy supply for their local government area and regions, including a population of around 24 million people as of 2016.[22] At least 74 regions have already reached 100% renewable electricity, and there are numerous examples of the 100% renewable energy target being exceeded, like Prignitz in Brandenburg, with 248% renewable energy.[23]

In the US, four cities are already entirely powered by renewable electricity: Aspen, Colorado; Burlington, Vermont; Columbia, Maryland; and Greensburg, Kansas; and a dozen more have made commitments to reach 100% renewables in the next 15 to 20 years, with many others considering similar plans.[24]

However, it is not only countries and cities transitioning to a cleaner energy future. Many organisations have analysed their situations and found that implementing their climate change reduction actions is no longer as expensive as they had once thought.

They realise that committing to the ambitious goal of going fully renewable makes economic sense, increases their market reputation, and is sound risk management against volatile energy prices. These companies innovated, embraced the change, and embarked on their journey. And because they are addressing this issue early, they will be far ahead of their competition.

More and more organisations continue to step up. They range from big retail stores like IKEA, IT companies like Microsoft and Apple, and car companies like BMW to smaller organisations that are eager to make a difference.

Most of these organisations have signed up to the RE100 campaign,[25] which was launched in 2014 by a coalition of businesses and not-for-profits. Participating companies have, on average, already transitioned half their energy use to renewables, with several companies having hit their 100% renewable energy target.

The following table provides examples of companies that have already achieved their goal, in order of their achievement; organisations that have set themselves goals until 2020, or 2050; and ones that have set themselves the objective, but have not yet specified a target year.

Renewable energy targets			
Achieved	By 2020	By 2050	No date yet
Swiss Post (2008)	Infosys (by 2018)	Aviva (by 2025)	Amazon
Pearson (2012)	Autodesk	Bloomberg (by 2025)	H&M
Commerzbank (2013)	BMW Group	Nike (by 2025)	HP
KPN (2013)	BT Group	Unilever (by 2030)	M&S
Biogen (2014)	Coca-Cola	Adobe (by 2035)	Nestlé
Microsoft (2014)	Goldman Sachs	Mars (by 2040)	P&G
SAP (2014)	IKEA	J&J	Salesforce
Steelcase (2014)	ING		Starbucks
Green Brewery (2015)	La Poste		Tata Motors
Proximus (2015)	Novo Nordisk		Walmart
Voya Financial (2015)	Royal Philips		
Alstria (2016)	SGS		
Elopak (2016)	Swiss Re		
Land Securities (2016)	Sky		
Google (2017)	UBS		

Table 1: 100% renewable energy goals and target years for companies

Why I have written this book

I grew up in Austria, a country that is traditionally placed in the top third of the OECD's sustainability ranking. Austria generates more than half of its electricity from hydropower, a couple of coal- and gas-fired power plants and, increasingly, from 'new' renewables like wind, solar, and biomass.

In 2003, I moved to Australia. Australia is an incredible, stunningly beautiful country, but most of its electricity comes from coal-fired power plants. Despite our advanced economy, most states in Australia still follow the old energy model, mainly due to the abundance of coal reserves.

Despite our historical reliance on fossil fuels, some regions in Australia have already embraced renewable energy on a large scale, with the most notable states being Tasmania (99.9% renewable energy), the Australian Capital Territory (well on track to reach 100% renewables by 2020), and South Australia (41.3%).[26]

The contrast between progress in Austria versus that in Australia became evident to me when my family and I visited Austria after having been away for some time.

As we approached Vienna International Airport, I saw the familiar Schwechat oil refinery below, which covers around half the demand for oil products in Austria. However, on our way to my former village, I could not believe the many changes that had occurred in a few years.

There was an endless number of wind farms along the highway, and the houses were covered with solar panels. In my village, the sporting ground had been relocated and in its place was a biomass facility that now provides central hot water and heating for nearly everyone.

The water pumping station and the telecommunications tower were partially powered by ground-mounted solar panels. People who were not connected to the central biomass facility had swapped the fuel in their boilers from oil to pellets from the local sawmill.

Over dinner, we drank beer from an Austrian brewery that runs on 100% renewable energy. Moreover, amidst all this change, everyone acted as if this was the way things were naturally done nowadays.

As I pondered this transformation, Lower Austria – my home state and the biggest of Austria's nine provinces – announced that it now generated 100% of its electricity from renewable sources.[27]

I had seen with my own eyes what the new energy model looked like and knew then this must also be possible in Australia. Whilst Australia is much bigger than Austria, with a population more widely spread out than those in Europe, we have renewable resources in abundance, the incentive of high electricity prices, and a tech-savvy community ready to embrace the change.

Eventually, every organisation will wean itself off fossil fuels and convert to renewable energy – it is not a question of *if*, but *when*. I have made it my mission to help organisations with their journey to 100%

renewable energy and to enable the transition in the most efficient and cost-effective way.

Together with my business partner, Patrick Denvir, I developed a proven methodology that helps transform organisations from the current energy situation which, in many places, is often dominated by the consumption of fossil fuels, to a future where renewables meet all of the energy demand.

This book is a result of our work in this space and my desire to share what we know so that you can more readily overcome any barriers that might otherwise delay you from joining organisations already embarking on this forward-thinking journey.

Who should read this book

This book will be immensely useful to any person looking to develop an energy strategy for their organisation, in particular, senior managers of both private and public sector organisations that are considering or tasked with transitioning to 100% renewable energy. It will also be valuable to anyone wishing to learn how to develop an energy strategy.

It is important to note that the principles and strategies contained herein hold true for *any* sustainable energy strategy, whether the goal is 100% renewables or just a portion.

Getting the most out of this book

You may choose to read the book from front to back, or you may prefer to pick and choose the chapters that most resonate with you. Either way, it is a good idea to revisit certain chapters over time to refresh your knowledge and gain new ideas.

This book contains a great deal of material. However, some of it is background information and simply details a best-practice method to help you achieve your goal with minimum effort, the least frictions, and the fewest setbacks.

To help you to identify the key actions you need to take to reach your goal, a checklist is provided at the end of each chapter. I have also interwoven private and public sector case studies, so you can see how other organisations have implemented the strategies described.

Embrace the change and commit to take action

'We choose to go to the moon. We choose to go to the moon in this decade and do the other things, not because they are easy, but because they are hard, because that goal will serve to organize and measure the best of our energies and skills, because that challenge is one that we are willing to accept, one we are unwilling to postpone, and one which we intend to win, and the others, too.'

– John F. Kennedy

In 1962, when John F. Kennedy decided to put a man on the moon by the end of that decade, the US had not yet even put a man into space. Kennedy knew he needed to inspire the American people and wake the imagination and resourcefulness of the country to win the space race against Russia.

Short-term goals are important to keep us moving forward. However, we need a long-term goal, an overall vision that is big and inspiring. Once Kennedy provided the vision, companies started to innovate, more resources were committed and, finally, in 1969, in a breathtaking moment, Neil Armstrong set foot on the moon.

The past decade has set the wheels in motion for a global energy transition to renewables. Change and disruption are happening fast.

As I write this Introduction it is 4 November 2016 and the Paris Agreement of December 2015 comes into force, legislating that the world must reach zero emissions by 2050.

The coming decades will belong to those organisations that envision achieving zero emissions and act to implement the necessary steps now. Instead of standing by the sidelines and watching everyone else embark on the journey to 100% renewable energy, you can place your organisation at the forefront and transition to a low-carbon economy.

Help change our current twentieth-century energy system to a sustainable one that is better for our economy, our environment, our society, our health, our climate, and your financial bottom line. This book will support your efforts to uncover your organisation's ingenuity, creativity, talents, and skills with a big, bold vision to transition to 100% renewable energy.

I would love to hear about your journey to reach 100%. Please share your stories with me at www.barbaraalbert.com.au.

The Four Steps to Achieve 100% Renewable Energy

FOUR STEPS

Figure 1: The Four-Step Methodology

Transitioning to 100% renewable energy is a new sustainability mega-trend, yet it is not always easy to do in the most cost-effective way and in the best interests of your business. Implementing the projects may take many years and, for most companies, the learning curve will be steep.

If your organisation has just committed to the target of 100% renewables, you are probably worried about the technical feasibility and the cost implications for your business. You might also wonder what options you have and how far out in the future you should set the target. You may be concerned about how the transformation will be financed and how long the payback periods will be. You might be unsure about how to get started, given you are also busy running your company.

Likewise, your customers or shareholders may pressure your organisation to reveal your long-term strategy for renewables and reducing your energy consumption. Perhaps your competitors are already moving down the road to 100% renewable energy sources and, by delaying action, you risk being caught 'on the back foot'. Can you afford, financially and reputationally, to be left behind?

On the other hand, you might be personally convinced that committing to 100% renewable energy is the only way to radically decrease carbon emissions, but unsure whether key organisational stakeholders, like executive management or asset owners, will support you in this endeavour. How do you present these ideas clearly and succinctly, and obtain executive management support?

The sheer breadth of a 100% renewable energy target can feel overwhelming, but it doesn't have to be. When you follow a proven methodology, such as the one provided here, the task becomes much easier. My practical four-step method is designed to enable your organisation to cost-effectively power your operations with renewable energy and energy efficiency options that are technically feasible and will earn the buy-in of your stakeholders.

The four steps have been created from my experience with organisations just like yours. I can say 'just like yours' because the steps are a process that has been successfully implemented across a range of sectors and business sizes.

Let's take a closer look at what these four steps represent.

Step 1 – Lead

Commit to a target of 100% renewable energy

- Appreciate the factors that drive businesses to set themselves such an ambitious objective.
- Define a goal that works for you.
- Understand potential renewable and energy-efficiency technologies, as well as battery storage and other future opportunities.

Step 2 – Plan

The more you plan, the easier it is to implement

- Analyse your current energy situation, and the potential for renewable energy technology to meet your projected energy demand.
- Prioritise your energy opportunities and pick suitable implementation, delivery, and financing options.
- Engage your stakeholders to get their support and package your preferred energy opportunities in a pathway.

Step 3 – Implement

Act on your identified opportunities

- Refine your business cases and get them market ready.
- Manage your project risks and organisational change, and find the right implementation partners.
- Monitor your progress towards your 100% target and revise the plan based on new information.

Step 4 – Succeed

Achieve your 100% renewable energy target

- When implementation is complete and you, or one of your partner organisations, operates and maintains your energy assets.
- Share your success and challenges with others.
- Celebrate your achievements.

Step 1 – Lead

If you think big, and set ambitious goals, then your achievement of them will be big as well. Goals channel your organisation's efforts and behaviour in a particular direction. Once you have a clear vision, you begin to narrow your attention and efforts to activities related to the goal, and start moving towards it.

The new standard for climate change leadership is 100% renewable energy, but is this an appropriate goal for your organisation? What does it mean to commit to 100% renewable energy? Must you be 100% renewable right away? Or can this be a long-term, staged target?

How is it possible to achieve the target, and what energy technologies get you there?

Step 1 – Lead, explores current sustainability market drivers and what makes organisations adopt such an ambitious target.

You set the scope of what is covered under your 100% goal and select an appropriate target year. Here, we examine renewable energy technologies and energy efficiency options for your electricity, heating and transport needs, as well as the opportunities battery storage and other emerging trends provide.

'Establishing a 100 per cent renewable energy goal helps us better serve society by reducing environmental impact. The pursuit of renewable energy benefits our customers and communities through cleaner air while strengthening our business through lower and more stable energy costs.'
– Mary Barra, Chairman and CEO, GM

Chapter 1

Committing to 100% renewable energy

'We are convinced this is good for business, this is not about greenwashing. This is about locking in prices for us in the long term. Increasingly, renewable energy is the lowest cost option. Our founders are convinced climate change is a real, immediate threat, so we have to do our part.'
Marc Oman, EU energy lead at Google

Moving to 100% renewable energy has many benefits to society: cleaner air and water, healthier communities, higher energy independence, local investment and employment opportunities, greater shared local ownership of the energy system, and a sensible option to help counter the problem of climate change.

However, not only will our society benefit from transitioning to 100% renewable energy but also your organisation. The advantages of 100% renewable energy for your business are locked in long-term energy costs, more control over your energy supply, greater energy security, a steady and increasingly lucrative return on investment, a strong reputation in the market, a more engaged workforce, and the reassurance that your organisation is doing its bit, contributing to a sustainable future.

When I analysed why organisations commit to 100% renewable energy, five major drivers became apparent, as shown below:

Figure 2: The five major drivers for committing to 100% renewable energy

Let's investigate these in more detail and see which ones resonate with you. They may even help you to build a business case as to precisely why your organisation should move down this path.

1. Meeting sustainability goals

Boards of directors and senior management teams are increasingly focused on corporate social responsibility performance, which is reflected in organisational sustainability-related targets and reporting.

The top performing organisations have set public sustainability goals for more than 20 years, but it is only since the 2000s that setting goals has become standard practice. According to the Ceres 'Power Forward' report,[28] 59% of the Fortune 100 and nearly two-thirds of the Global 100 have set carbon emissions reduction commitments, renewable energy commitments or both.

On 1 January 2016, the 17 Sustainable Development Goals (SDGs) of the 2030 Agenda for Sustainable Development officially came into force.[29] As I mentioned in the introduction, these SDGs are a set of global goals adopted by the 193 UN member states, of which goal seven, 'Affordable and clean energy',[30] and goal 13, 'Climate action',[31] directly relate to renewable energy.

While it is governments that sign up to the SDGs, businesses are needed to help achieve those goals, which is what the UN Global Compact, the world's largest corporate sustainability initiative,[32] wants to achieve. According to research from PricewaterhouseCoopers,[33] 70% of businesses plan to embed the SDGs within five years.

Implementing energy efficiency and renewable energy initiatives helps achieve your sustainability goals. Previously, the implementation of renewable projects was expensive, and came with a long payback period, which resulted in them being parked for future implementation. The time is now right to revisit these opportunities.

With the recent price falls, switching to renewable energy is a cost-effective way to meet climate change mitigation targets. Renewables enable your organisation to set more stringent carbon reduction goals, with 100% renewable energy being the most ambitious one.

Committing to 100% is the new standard in climate change leadership

Many organisations have greenhouse gas reduction targets in place and, over the years, yours likely has worked on implementing 'low-hanging fruit' opportunities, such as increasing energy efficiency. With all your successes, you might be wondering what should be next on your agenda and how much further you can reduce your carbon footprint.

Recently, the expectations in the marketplace for transparency, disclosure, and measuring progress towards goals have climbed. Standards for targets have also consistently risen, making it harder for organisations to stand out unless they commit to exceptionally difficult-to-reach ones. The new standard in climate change leadership is to commit to 100% renewable energy.

Local government sustainability goals

Committing to 100% renewable energy is the new standard for businesses, but increasingly it is also becoming the new standard for cities and towns. Sometimes this is led by the executive management within these organisations; in other instances, it is driven by communities and community leaders.

Cities have the opportunity to participate in a number of programs, which has resulted in many of them setting sustainability goals. Current programs for cities include the 'CDP for Cities', the 'Compact of Mayors', 'C40', and the 'Carbon Neutral Cities Alliance'.

C40 is a network of the world's megacities committed to mitigating climate change. The CDP for Cities works with more than 300 cities to address their carbon emissions. The Compact of Mayors is the world's largest coalition of city leaders addressing climate change by pledging to reduce their greenhouse gas emissions, tracking their progress, and preparing for the impacts of climate change. The Carbon Neutral Cities Alliance is a collaboration of international cities committed to achieving aggressive long-term carbon reduction goals.

While the above-mentioned programs are typically adopted by larger cities, most smaller local governments have participated in the Cities for Climate Protection (CCP) program. The CCP program was sponsored by the International Council for Local Environmental Initiatives (ICLEI), which wanted to drive the development of greenhouse gas action strategies for corporate as well as community greenhouse gas emissions.

Local governments participating in the CCP program started to actively monitor, manage, and report on their carbon emissions on the basis that one cannot manage what one does not measure. They were also looking for ways to reduce their carbon footprint and thus their energy consumption.

In my work with local governments, I have found that the Councils that commit to a 100% goal have been early participants in this program.

They see making the ambitious commitment as a logical extension of the original program and use this goal to help meet their carbon reduction commitments.

However, it is not just programs like the CCP that drive the sustainability agenda of local governments. Whilst boards of directors influence businesses, the direction of municipalities is affected by their elected councils which, in turn, are influenced by their residents.

Lismore City Council, in Australia, for example, undertook an extensive consultation of their residents in 2012 and 2013 to determine where it should head in the medium and long terms. Council was not actively looking for an environmental mandate, but the results of the community engagement were astounding: the residents wanted council to be 'a model of sustainability'.

In support of the community's vision, council adopted a target to 'self-generate all of Lismore Council's electricity needs from renewable sources by 2023', a goal only 10 years distant at that time. Lismore Council developed a strategy and action plan to help it achieve its objective and is now well on the path to implementing all the identified projects.

2. Changed market conditions

Customers have become more aware of sustainability issues. The pressure to be more environmentally responsible comes from all kinds of stakeholders, like customers who want lower prices, and better products that have been produced more sustainably, shareholder groups that want to reduce their risk, not-for-profits that want to see your carbon emissions decline, your peers who are lifting the bar, and your employees.

Shifting customer demand

Volatile energy prices, polluted ecosystems, a growing awareness of climate change and the geopolitical costs associated with fossil fuels alter the way people relate to energy. More and more, people think about where their electricity comes from, how they use it, and what impact all of this has on our planet and people.

Consumers have become much more aware of the damage organisations can cause and increasingly want to deal with businesses that produce their products and services with clean, renewable energy.

According to the Global Consumer Wind Study,[34] which surveyed more than 24,000 consumers in 20 countries, 74% of consumers said they would have a more positive perception of a brand if wind energy were the primary energy source used in its production, and 49% of consumers expressed willingness to pay more for products made with renewable energy. This new demand from customers means changing market conditions, which your organisation will have to respond to.

If your organisation serves other businesses, rather than end consumers, you might face a changed landscape as well. Organisations that lead in sustainability have begun to look beyond their own four walls to tackle today's environmental challenges. They gather more information on the sustainability commitments and achievements of their key supply chain partners, one of which may be your organisation.

Apple, which has committed to 100% renewable energy, is working with its suppliers to help reduce its carbon footprint. In 2016, Lens Technology, a major supplier, agreed to run its Apple operations entirely on renewable energy.[35]

Being in a position to respond with details about your energy consumption and your renewable power generation is imperative and improves your supplier ranking. However, imagine the impression you will make on your customer if you target 100% renewable energy, and can show a gradual transition to that goal.

Also, the fact that your client asks you for your environmental credentials indicates they are sustainability leaders. How likely might your customer commit be to 100% renewable energy? And if they do, how likely might they be to expect a similar commitment from you, in the future? Are they also looking at your competitors and their approach to sustainable outcomes?

Peer pressure

Once a few companies in your industry sector make a commitment to go to 100%, you will wonder whether it is better for you to make the same move. For instance, most big names in the retail space, like Walmart, IKEA, Nike, H&M, Marks & Spencer, Mars, Nestlé, Procter & Gamble, Starbucks, and Unilever have already signed on to a 100% renewable energy goal. What does this mean for all remaining retailers?

In Australia, National Bank Australia (NAB) declared a goal of carbon neutrality. This was rapidly matched by ANZ and Westpac, two more of Australia's largest banks.

Another example is the commitment of cities to go 100% renewable. In 2014, only a handful of cities in Australia committed to the goal, most of which were located in areas where residents strongly support sustainability initiatives. Since then, many more cities have made this same commitment, and it is only a matter of time before many others follow suit.

Investor preference

Changing market attitudes also affect investment decisions. People and organisations with a strong ethical agenda are divesting funds from businesses powered by (or producing) fossil fuels. As part of their risk management, potential investors also want to know how companies cope with the energy market's growing volatility. More and more people and organisations are investing in renewable energy, not only for ethical reasons but also because of good investment returns.

Grassroots climate movement organisations like 350.org lobby for oil, gas, and coal to remain in the ground, and actively encourage organisations to divest from fossil fuels. To date, nearly 700 institutions globally, representing more than US$5 trillion in assets under management, have made some form of commitment to divest from fossil fuels.[36]

Examples of organisations that have already divested from fossil fuels include universities, like the Australian National University, Stanford and Oxford; organisations representing millions of people, like the World Council of Churches; trusts, like the Rockefeller Family Fund; and municipalities like Paris, France; Canberra; and Newcastle, in Australia, which incidentally is home to the world's biggest coal port. In Australia alone, 16 local governments have committed to divest.

Apart from shareholders divesting from fossil fuels, other investors increasingly assess companies on their downside risk from greenhouse gas emissions and their upside potential from reducing them, in light of current and potential future government regulations.

The CDP (formerly known as the Carbon Disclosure Program) has requested information on the risks and opportunities of climate change from the world's largest companies on behalf of more than 80 institutional investor signatories since 2003.

Not only do they send surveys to publicly listed companies to disclose their climate change risks and what they are doing about them but also request this information from large cities. The respondents need to state their renewable energy goal and progress towards it, with the results collated in annual reports that anyone can look up.

There are also other shareholder groups that request companies to report on their emissions profile, including information about the percentage of renewable energy. Examples of these groups include Asia Investor Group on Climate Change, Investor Group on Climate Change, Institutional Investors Group on Climate Change, Investor Network on Climate Risk, and the Asset Owners Disclosure Group, amongst others.

Some of these shareholder groups have formed an alliance called 'Aiming for A', which forces the boards of large, listed organisations to put a resolution to a shareholder vote to compel a company to improve its emissions management, which includes the transition to renewable energy.

Public scrutiny and work by not-for-profits

Public expectations and the lobbying work of not-for-profits comprise another reason organisations commit to 100% renewable energy. Sectors that consume a lot of energy, like heavy industry, or cloud computing businesses with power-hungry data centres, like Amazon or Google, are frequently the centre of attention.

In 2014, Greenpeace came out with the 'Clicking Clean' report on how cloud companies create the 'Green Internet'.[37] It was a naming and shaming report that listed all major digital leadership companies, including Apple, Google, and Facebook.

Companies were scored in the following categories: percentage of renewable energy powering the organisation, energy transparency, renewable energy commitment and siting policy, energy efficiency and mitigation, renewable energy deployment, and advocacy.

The key findings of the report were that companies like Apple, Facebook, Google, and Salesforce had committed to a goal of 100% renewable energy and that Google maintained its leadership in building a renewably powered Internet.

Apple and Facebook were applauded for their transparency with energy-specific data. Previously, companies within the sector withheld information due to competitive concerns.

However, the report publicly shamed Amazon and Twitter, claiming that 'Amazon Web Services remains among the dirtiest and least transparent companies in the sector, far behind its major competitors, with zero reporting of its energy or environmental footprint to any source or stakeholder. Twitter lags in many of the same areas.'

The public shaming by Greenpeace worked, because a few months after the report came out, Amazon committed to using 100% renewable energy for its cloud operations.

The Greenpeace report targeted leading cloud-based companies, but there are also examples of not-for-profits targeting cities to become greener. The Sierra Club, with its #ReadyFor100 campaign, works to get cities to commit to 100% renewable energy.

With 2.4 million members and supporters, the group wants to harness its grassroots power to urge mayors to follow the lead of global cities such as San Diego, Paris, Sydney, and Vancouver, all of which have made 100% renewable energy or net zero emissions commitments.

There are also many not-for-profits working to increase corporate demand for renewable energy. One such example is RE100, a collaborative, global initiative of influential businesses committed to

100% renewable electricity. Companies joining RE100 are encouraged to set a public goal to procure 100% of their electricity from renewable sources of energy by a specified year. Participating businesses have, on average, already transitioned half of their electricity use to renewables, and several companies have hit the target, according to a report released by RE100 in March 2016.[38]

RE100 was launched in September 2014, with 12 partner organisations. At the start of 2016, there were 50 partner companies in the RE100 campaign. In December 2016, there were more than 80 companies, with more continuing to sign up. Current partners include Adobe, BMW, Coca-Cola, Goldman Sachs, Google, ING, Johnson & Johnson, Mars, Nike, Procter & Gamble, SAP, Starbucks, UBS, and Walmart.

Once your organisation has joined not-for-profits, like RE100 or the Rocky Mountain Institute and its Business Renewables Center (BRC), you will enjoy the positive publicity that comes with being part of these campaigns and groups. The BRC helps organisations purchase renewable energy at competitive electricity rates by aggregating demand from several companies that operate in a specific location.

All this work by not-for-profits will result in more organisations and cities committing to 100% renewable energy, which will snowball into others following their example. If this trend continues, which is likely, powering companies and cities with 100% renewable energy will soon be seen as the standard way to do things.

Employee engagement

According to a survey performed by *Corporate Responsibility Magazine*,[39] corporate reputation has a significant impact on the attractiveness and expense of talent acquisition and retention. Nearly three-quarters (74%) felt it important to work for an organisation led by a CEO whose priorities include corporate responsibility and/or environmental issues.

Committing, and gradually moving closer, to 100% renewable energy, especially where employees feel a sense of ownership, achieves greater engagement levels. Employees engaged in their work are more motivated than disengaged staff, and focus better on achieving business goals.

3. Cost reduction

How many times in recent years has your company had to increase the budget line item for electricity? In Australia, business electricity prices increased 82% over the period spanning 2007–08 and 2013–14, compared with a 13% increase in the producer price index.

Energy is an important business input and many organisations feel the impact of rising prices. The cost of energy is projected to continue to rise due to infrastructure costs to transmit and distribute energy, commodity prices, exchange rates, and geopolitical factors.

Rising costs can be counteracted by being more efficient in your operations and implementing renewables. The cost for solar, wind, and LED lighting has dropped notably in the past 10 years. The fuel for most renewables is free, whereas the price for non-renewables will invariably rise. Having long-term price certainty has tremendous value for organisations with large energy consumption.

Great return on investment

There are many energy efficiency and renewable energy projects that come with a financial return much greater than your lending rate, or other investments your organisation might make. Nowadays, due to factors like improvements in technology, and more specialist suppliers and producers in the marketplace, it is easy for sustainability projects to compete financially with other projects.

Locking in energy prices for the long term

For the better part of the last ten years, electricity prices have only had one trend: up. Many of my clients have tried to counteract the trend of rising prices and have spent a lot of time and resources on reducing their electricity consumption through efficiency measures.

An example of how the energy consumption of one of my local government clients has trended down, whereas the costs have increased, is shown in Figure 3. The bars illustrate the energy consumption, whereas the line demonstrates the costs (not inflation-adjusted).

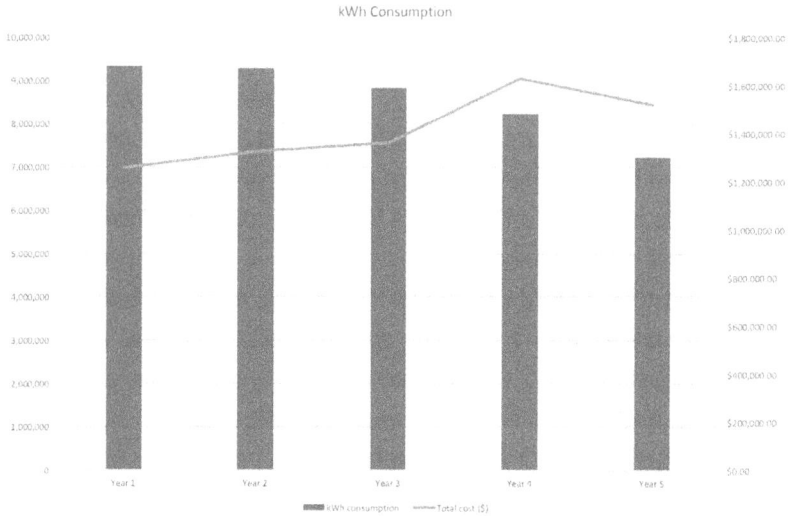

Figure 3: Example of reduced energy consumption with rising electricity prices

The cost line has gone up, rather than down, due to the increased price of grid-supplied electricity. The cost increases would have been even higher without the energy savings measures that my client implemented.

Many boards of organisations I work with are concerned about the volatility of energy prices. Only recently, I was consulting to a business that faced a $3 million increase in its energy bills in the coming financial year. The board decided to investigate how the deployment of onsite renewable energy could be maximised to mitigate this price rise. The board knew that once it set up the renewables infrastructure, the business would not have to pay for any fuel.

Another option this organisation could consider is to enter into long-term power purchase agreements (PPAs) for offsite renewable energy projects. With PPAs, they might find that they can lock in their prices for 10, 15, or 20 years.

The volatility of energy prices can also be felt at the bowser. Ten years ago, oil-based fuels in Australia were fairly cheap, before we went through a period of rising fuel costs, and prices climbed by more than 50%. Recently, with extremely low oil prices, the cost of fuel has gone back to almost the level of 10 years ago.

Even with oil prices at record lows, not all price falls are passed along the whole supply chain to the end consumer. Rather than being dependent on foreign pricing policy, you can consider a long-term plan to change your fleet's fuel source to electricity generated from renewable sources, like the sun, or wind, with no associated fuel costs, thus locking in your energy costs for the long term.

4. Managing risk

You do not want your organisation to be exposed to unforeseen risks. Transitioning to renewables will help you risk-manage energy vulnerability. It will also equip you better for a carbon-constrained future in light of more climate change-related regulations being put in.

Energy vulnerability

Energy vulnerability comes in the form of grid blackouts or being dependent on fuel sources that are controlled by others. If the grid is unavailable, a blackout can seriously affect organisations, especially critical infrastructure like banks or hospitals.

There are numerous examples of bushfires in Australia taking down the grid, which can take days or even weeks to fix. In September 2016, a storm caused a statewide blackout in South Australia. In December 2015, Ukraine suffered the first confirmed cyber security attack,[40] which took down a big part of the national grid.

It is expected that the growing impacts of climate change, severe weather events and hacker attacks will increasingly affect the reliability of the grid in countries that have not invested in the stability of the grid, or where companies are located on the fringe of the grid. Some organisations may find it important to get prepared for power outages. One of the goals for IKEA, for instance, is to 'strive for resource and energy independence', which will give it greater control over its energy use and energy sources.

While previously organisations relied on diesel generators for power outages, they are now evaluating switching to renewables, in conjunction with battery storage. The advantages of using renewables rather than diesel is that there is a local supply of renewables and that the available quantity of energy is not capped to the litres of diesel that are currently available.

Energy vulnerability affects your stationary operations, but it also affects your transport needs. If you run your fleet on petrol and diesel, you are buying fuels that are controlled by others. The supply and price of these fuels are set by oil-producing countries and if they decide to reduce the production, what will happen to the cost of those fuels?

Rather than relying on imported oil, you can strive to switch to renewables that are available locally, which results in money staying in the region, more political stability, air quality improvements, local employment benefits and more energy security for your organisation.

Regulatory risk

As concerns mount over global climate change and as there is more broad public support for cuts to emissions, governments will intensify regulations to move to a low carbon economy and compliance costs for carbon-intensive business models will go up.

Regulatory costs can come in the form of decreased or removed subsidies for fossil fuels. They can come in the form of a price on carbon, examples of which are a carbon tax or an emissions trading scheme. The impact of these rules on industries varies from inconsequential to game changing, but moving to a 100% energy model will most likely help drive your compliance costs down.

In Australia, we have a turbulent history of carbon pricing. Back in 1997, Australia was one of only two industrialised nations[41] not to ratify the Kyoto Protocol (reversed in 2007, when we did sign it). In 2012, a fixed

Carbon Pricing Mechanism (CPM) was enacted, which operated for two years, after which it was abolished. It was replaced with a reverse-auction scheme meant to encourage organisations to voluntarily reduce their carbon emissions and receive money for doing so.

Apart from a price on carbon, policy can also come in the form of increasing national laws and regulations that affect how you plan for or manage energy. Examples are building codes; special requirements for getting development approvals, like energy efficiency requirements; laws that force you to calculate the payback periods for energy efficiency and renewable energy projects; or fuel efficiency standards for vehicles.

Already three-quarters of all new cars sold each year globally are regulated under some form of carbon dioxide emissions standard. There are also numerous examples of governments in Europe providing both mandatory and voluntary standards for the energy performance of buildings.

As per the Paris Agreement, all our buildings need to be zero net emissions from mid-century onwards, so you need to start planning for aggressive energy efficiency and renewable energy generation now.

Building regulations are already starting to extend to onsite renewable power generation, not just energy efficiency measures. Some governments in the US, for instance, have already started to mandate solar PV on roofs.

In San Francisco, new legislation requires solar PV systems, solar water heating systems or a combination of the two for new commercial and residential buildings of up to 10 storeys. Incidentally, San Francisco has a goal to use 100% renewable energy by 2025.

California's *Long Range Strategic Plan* sets forth a goal for all commercial buildings to be built to zero net energy standards by 2030. Another example is Israel where most buildings use solar hot water heaters, thanks to a mandate. If this model is successful, there is a strong likelihood that other states and countries will follow.

Rather than waiting for these regulatory risks to affect your organisation, it is better to transform the challenge into opportunities by putting effective plans in place now.

Setting yourself the goal of 100% renewable energy will force you to be much more proactive in your decarbonisation efforts than you otherwise would be and it will help you avoid putting suboptimal solutions in place that have not been planned for.

5. Leadership and innovation

In December 2015, the Paris Agreement at COP 21 set global sights on a more ambitious 1.5-degree Celsius path, which cannot be reached without a massive contribution from renewables. For the first time, the international community has committed to net zero greenhouse gas emissions in the second half of this century.[42]

This successful outcome has driven investment in renewables even further forward and has increased business confidence in the rollout of renewable energy. As the global economy undergoes a deep transformation, your business can be recognised as a leader amongst your peers by taking climate action.

The so-called 'Paris effect'[43] has already caused many organisations to step up their commitments. The following quote is from Paul Polman, CEO of Unilever, who spoke at the COP21 summit. I highly recommend you watch his speech on YouTube.[44]

> The consequences of this agreement go far beyond the actions of governments. They will be felt in banks, stock exchanges, boardrooms and research centres as the world absorbs the fact that we are embarking on an unprecedented project to decarbonise the global economy. This realisation will unlock trillions of dollars and the immense creativity and innovation of the private sector who will rise to the challenge in a way that will avert the worst effects of climate change.

Public recognition and awards

Moving to 100% will make your organisation a low carbon energy leader and potentially a net exporter of renewable energy. Having such an ambitious goal is seen as proof that you prioritise our people's and planet's health and well-being over the lure of special interests that tend to earn organisations a lot of positive media coverage.

I recently had a discussion with Andreas, the brewery master of the 'Green Brewery', the largest and most sustainable brewery in Austria. Over the span of 10 years, it has implemented industry-changing projects that have led the company to 100% renewable energy. Its achievement has inspired other organisations to follow the example, and visitors come from all over the world to see it firsthand and learn from the brewery's experience. (I detail the brewery's pathway in Chapter 11.) The brewery has also been featured positively in the news and receives many requests to share its experience at conferences and webinars.

Making the effort to switch to clean energy will also provide an opportunity to receive industry awards. As an example, one of my clients, Lismore City Council, won a prestigious industry award for the development of its 100% renewable energy plan and was selected as one of only two case studies representing Australia at the COP21 conference.

Competitive advantage, innovation and being a leader

Many organisations are driven by being leaders in their field and are continually setting themselves high standards with short, medium and long-term goals. Leading organisations have a vision and purpose that transcends the mere drive for profits. They have an ability to see the big picture and are willing to take a calculated, data-informed risk. Such organisations set a high bar and motivate others to do the same.

Every organisation will have its own priorities and sector-specific issues to balance, but to stay a leader they all need to innovate. Organisations that fail to innovate run the risk of trailing behind competitors, losing key staff, or simply operating inefficiently.

Organisations that are not mapping the path to 100% now might be caught on their technological back foot. Leaders and fast followers will stay cost competitive whereas slow adapters might suffer as a consequence of missing this trend.

Successful organisations not only respond to their current customer or organisational needs but also anticipate trends, which allows them to meet future demand rapidly and effectively.

IKEA, for instance, has a business interest in developing its expertise in solar PV systems. It is considering partnering with a solar energy firm to provide IKEA customers with an in-store consultation service, as well as installation, maintenance and monitoring services.

If your organisation is a leader and embraces innovation, then setting yourself a goal of 100% renewable energy is a natural extension of your leadership position.

Your checklist:

You may be able to delegate or outsource these tasks.

- ☐ Determine which of the drivers in this chapter relate the most to your organisation.
- ☐ If your organisation has not committed to 100% renewable energy yet, prepare a document and presentation to senior management. Show them the international and local context, as well as relevant examples of organisations that have already committed to or achieved the target.
- ☐ If your organisation has already committed to 100% renewable energy, you can use information in this chapter to further substantiate your organisational goal.

Chapter 2

Laying the groundwork

Before developing a strategy to achieve 100% renewable energy, it's helpful to know a few basic principles and definitions, starting with the difference between fossil fuels and renewable energy, what a goal of 100% renewables means, and the specific boundary suited to your particular situation.

The next step is to investigate what type of energy will be part of your commitment, which operations you will include, what your options are to express your goal, and what an appropriate timeline will be. We will also explore the advantages and disadvantages of renewable energy certificates/RECs relative to reaching your goal.

Fossil fuels versus renewable energy

Currently, most of our energy consumption worldwide is supplied by fossil fuel energy sources, primarily oil, gas, and coal. Fossil fuels formed over millions of years in the Earth's crust from decaying organic matter that used the sun's energy to convert carbon dioxide and water into molecules needed for growth.

When those fossil fuels are burnt, the energy stored in them is unlocked and the carbon is released back into the atmosphere in the form of carbon dioxide. Carbon dioxide and other greenhouse gases trap the sun's heat. The more greenhouse gases we add to our atmosphere, the more we heat up our planet, to dangerous levels.

Another problem with fossil fuels is that once they are consumed, they cannot be replenished. In Australia, 85% of electricity is produced by burning fossil fuels, whereas only 15% is currently produced from renewable energy sources.[45] It is not surprising that Australia uses that much fossil fuel: we have plenty stored underground and it is cheap to extract.

The world will not run out of natural gas and coal reserves anytime soon, and, in Australia, we are endowed with enough black coal to last another 100 years and enough brown coal to last another 465 years,[46] at current rates of extraction.

Renewable energy, on the other hand, is obtained from natural resources that can be continually replenished. The main sources of renewable energy are the sun, wind, water, geothermal (the earth's internal heat), and biomass (organic matter).

Renewable energy sources are accessible across broad geographical areas whereas fossil fuels like oil and gas are massed in far fewer locations. Most renewable energy is clean and there are no emissions associated with the operation of the technology.[47]

This book does not describe opportunities based on switching from a highly carbon-intensive to a lower carbon-intensive fuel. Whilst it is true that if you change from coal to natural gas, *fewer* greenhouse gases are released, but these greenhouse gases still contribute to climate change.

The goal here is to encourage you to switch to fuels that do not release any fossil carbon at all. Ideally, our remaining fossil fuels should be used only to establish the renewable infrastructure of the future.

What does a goal of 100% renewable energy mean?

There is no official definition (yet) of what a 100% renewable energy target means, but a lot of organisations interpret it to be when the amount of renewable energy produced is equal to or more than what is consumed.

An organisation is 100% renewable when its annual energy consumption is met, or exceeded, by an equal amount of renewable energy.

The renewable energy can be generated either on- or off-site. As long as the renewable energy supply meets or exceeds the energy consumption of your organisation, you have reached your 100% renewable energy goal.

People sometimes think that to be 100% renewable, the energy consumption must be matched in real time with a renewable energy supply. If this were the case, the technology, administration, and documentation challenges would be significant. It would also expose organisations to unwanted price fluctuations, as any shortfalls would have to be met by purchasing renewable energy in the short-term energy (spot) market.

Instead of matching current energy consumption with real-time renewable energy supply, we will use a concept called 'net use accounting', where your yearly energy consumption is balanced by an equal amount of renewable energy generation.

The grid plays a vital function in net use accounting. Where on-site generation of renewable energy (and storage) does not meet the site's demand, electricity is consumed from the grid. When on-site renewable energy generation is greater than the site's load (and any capacity from battery storage), excess electricity is exported to the grid. The grid is also essential for off-site generation.

Independence from the grid

Clients often ask if by switching to 100% renewables they can become independent of the grid. Transitioning to 100% renewable energy without being connected to the grid is expensive and hard to achieve. The renewable energy system would have to be oversized, and require a large amount of (battery) storage on-site, to account for variability in generation and consumption.

If, however, your organisation, or parts of it, are off-grid, and powered by diesel generators rather than grid-supplied electricity, a combination of renewables and storage might be feasible and cost effective.

Setting your goal's boundary

Now that we know what 100% renewable energy means, it is time to think about how to interpret it for your organisation. Where will you draw the boundary in terms of your organisation's location, your energy forms, and the renewable energy sources that are acceptable to you?

This section investigates these concepts in greater detail and looks at a few case studies on how organisations have interpreted and expressed their goal.

Organisational boundary

If your organisation is spread geographically over a wide area, you need to decide what the 100% renewable energy goal relates to. Will it cover all of your operations, or maybe one country, or maybe only one site? Novo Nordisk, for example, has set itself the target to 'use electricity from renewable sources at all its global production sites by 2020'.[48] It has made it clear that all its production sites will be covered.

Will your target extend to assets that are owned or controlled by you, or will you spread it out to other parties connected to your organisation?

Municipalities can set the target for just their local government operations or for the whole community. Organisations can focus on their operations or extend the boundary out to sources like the energy consumption of their employees.

Organisational preferences

In addition to setting a geographical boundary, you should consider what kinds of renewable energy technologies are acceptable as part of your pathway. For instance, if you represent a local government, would you accept the burning of landfill waste, which incorporates oil-based plastics, as a renewable technology? Would you accept sewage gas or biomass facilities and, if so, would they have to meet certain criteria?[49]

Would you rather balance all of your energy consumption with the purchase of renewable energy or would you rather produce your own renewable power? Lismore City Council originally set itself the goal to 'become 100% self-sufficient in electricity from renewable sources by 2023'. However, when I ran workshops with staff and the councillors, the goal was refined to something more ambitious: the Lismore council wanted to 'self-generate all of council's electricity needs from renewable sources by 2023'.[50]

There is a difference between 'becoming 100% self-sufficient in electricity' and wanting to 'self-generate all of council's electricity needs'. Lismore preferred not to purchase renewable energy from outside its local government area. It felt that to show true leadership, the electricity should be generated within its geographical boundary.

Drawing the boundary around your energy forms

SAP, a German software corporation, has committed to achieve net zero carbon emissions by 2050 and eventually source 100% of its electricity from renewable sources.[51] Procter and Gamble have a short-term goal to 'source 30% of its energy from renewable energy by 2020', and a 'long-term goal to power all its plants with 100% renewable energy'.[52]

Did you notice how sometimes the term 'energy' and at other times the term 'electricity' is used? There is a big difference between these two.

For most organisations, 'energy' encompasses not only electricity but also stationary fuel use, as well as transport fuels. Examples of stationary fuels are natural gas, diesel for generators, LPG for forklifts and, in some instances, coal for boilers. Examples of transport fuels include the diesel, petrol, and LPG that power your fleet.

The discussion about 100% renewables tends to centre on electricity, but that is only one part of the energy picture. In Australia, fossil fuel-based electricity accounts for roughly 28% of its total net energy consumption.[53] Transport accounts for 27%, manufacturing for 19%, and mining for 9%.

To make electricity, stationary fuel use, and transport energy renewable, it is necessary to transition them all away from fossil fuels to renewable sources.

In the current market conditions, switching your electricity supply to renewables is the easiest goal to achieve. All you have to do is offset your electricity consumption with an equal amount of renewable energy production.

Tackling stationary fuels, especially in the manufacturing sector where there is a need for energy-intensive heating or drying, is more challenging, but doable.

There are established and technically feasible renewable energy solutions, like bioenergy, solar thermal, and heat pumps, which can substitute for stationary fossil fuels. This is most cost effective for temperature requirements under 250 degrees Celsius, and there are numerous examples around the world of where organisations have successfully implemented renewables.

Adding transport energy to the boundary will truly make your organisation 100% renewable, but you might want to allow yourself more time to achieve this goal. Your options are to either switch to biofuels or to electrify transport and power it by generating renewable energy.

You will be able to electrify your light fleet over the next years, but electrifying your heavy vehicles, or changing them to biodiesel, might take longer, until these technologies become more widely available. You will also have to think about whether 'transport' energy only relates to the fleet owned or controlled by your organisation, or whether it extends to the vehicles that employees use to commute as well.

The San Diego City Council, for instance, has a goal to shift half its fleet to electric by 2020 and increase the number of zero-emission vehicles in the fleet to 90% by 2035. It also wants to achieve 100% renewable energy for its electricity supply by 2035.[54]

You can use the following simple diagram to draw your own boundary for the energy forms your goal should cover. When you undertake this activity, bear in mind how much of each energy source contributes to your overall energy consumption.

If you are a trucking company and you exclude transport fuels, you might open yourself to criticism. Likewise, if you are a metals manufacturer, with significant gas usage, think carefully before excluding stationary fuel consumption in your 100% renewable energy target.

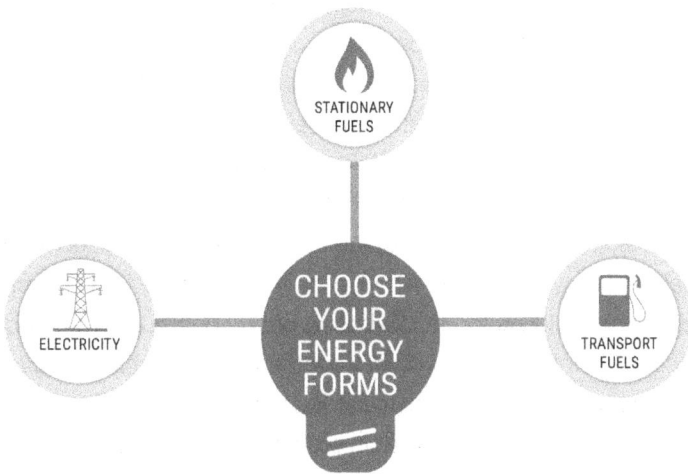

Figure 4: Determining the boundary for your goal

Timing your goal

Deciding to power your organisation with 100% renewable energy in the most cost-effective way is a long-term journey that may well take years to achieve. To increase your chances of achieving the target, setting a deadline for your goal is recommended. This adds a sense of urgency, which pushes your organisation into action. It transforms an abstract wish into a concrete objective.

Clients often ask me how far out in the future they should set their end goal. The answer is to budget enough time to complete the internal planning work, allocate resources with accountabilities, train and inform staff, obtain the necessary approvals, get the funds allotted, and allow for the majority in the organisation to become ready and supportive of the projects.

It also depends on what renewable technologies suit your circumstances, and what their maturity and cost curves look like. If a particular technology is not yet mature or cost effective, maybe it will be in five years' time. If you work out projected costs for renewable technologies, you will have an educated view of when in the future you should set yourself the end goal.

Several leading organisations that started their sustainability journey a decade ago have already arrived at their target. Some companies, like Microsoft, or municipalities, like Wildpoldsried in Germany, or Aspen in the USA, are already powered by 100% renewable energy.

However, the bulk of organisations that have committed to 100% renewable energy have yet to achieve the target. Apple, for instance, has already reached 100% for its US-based operations but are still working to transition their locations in other countries to renewable energy.

If you are not quite sure when to set your target year, you can undertake the work during *Step 2 – Plan* first, in order to make an informed decision before you announce your plans to the market.

Interim targets

One local government client, Coffs Harbour City Council,[55] chose an interim target in addition to the final one to help keep it on track:

- 50% corporate GHG emissions reduction from 2010 levels by 2025
- 100% renewable energy for council's facilities by 2030.

Another example of staggered targets is Unilever. Unilever aims to be 'carbon positive' in its operations by 2030.[56] To achieve this, the company is committed to source 100% of its energy across its operations from renewables by 2030, to source all grid-purchased electricity from renewables by 2020, and to eliminate coal from its energy mix by 2020.

In order to achieve its carbon-positive target by 2030, Unilever intends to directly generate more renewable energy than it consumes, and to make the surplus available to local markets and communities.

Define a goal that works for you

There are several options available to define a goal and a boundary that works for you. Management can make a top-down decision as to what should and shouldn't be included, or someone can be tasked to make and present a recommendation.

You can engage consultants to help you make the best decision, or you can hold a series of meetings or a workshop with key decision makers, or a combination of these options.

The end result will be a 100% renewable energy goal definition tailored to your needs and capabilities, one that expresses your organisational vision for a sustainable future. Once you have this, you can announce the goal through your usual communication channels.

To sell or not to sell renewable energy certificates

Renewable energy certificates (RECs) were created to spur the development of renewable energy generation through a market-based mechanism of supply and demand. An REC embodies the environmental attributes of renewable energy generation and has a financial value attached to it, which fluctuates depending on market conditions.[57]

Once electricity from renewable sources enters the grid, it mixes with electrons from multiple sources, like coal-fired power plants, and becomes indistinguishable. To track renewable energy, RECs are assigned for every MW hour created from renewables.

Each REC is assigned a unique number to track the ownership of the environmental (and social) benefits of the renewable energy. They can be traded separately from the underlying electricity.

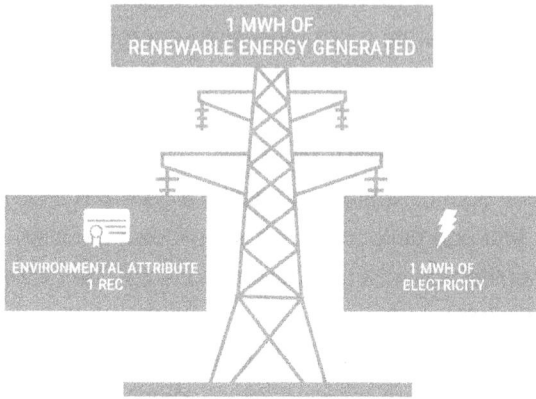

Figure 5: Renewable electricity and the generation of RECs

The party that owns the REC owns the claim to that MW hour of renewable energy.

If you install a renewable energy system, you may be eligible to receive renewable energy certificates. Most organisations sell their RECs and receive money in return, which reduces the payback period for the installation of a renewable energy project.

However, if you sell your RECs to another party, only the party that purchased them can claim the environmental benefits from your renewable energy. If you also claimed the renewable energy generation, there would be double counting. Organisations that are not aware of this issue could sell their RECs and mistakenly still claim the renewable energy generation. Whilst not illegal, it risks making them look bad, even dishonest.

In Australia, for instance, local governments have to report their environmental performance in the 'State of the Environment' report. Part of that report is a section on greenhouse gas emissions. Many councils claim an emissions reduction due to their solar installations, even if they sold the RECs.

Other councils, aware of the double-counting issue, do not claim the emission reduction, even though they have also installed solar panels. This leads to some looking comparatively better than others.

If you decide to sell your RECs, you want to avoid exposing yourself to potential reputational damage in the future, especially in the year you intend to meet your renewable energy goal. In the truest sense of a claim of '100% renewable energy', you must withdraw and retire your eligible RECs, which means you cannot sell them.

However, selling RECs helps make renewable energy a more attractive investment, so one option to consider is to sell your RECs in the years before your goal of 100% renewable energy is realised, and then retire the RECs once you reach your target year.

Another option is to re-word the goal and reframe the claim. Rather than state '100% renewable', you can claim that your annual energy consumption has been met by the 'generation of renewable energy'.

You can consider expressing your goal like IKEA did: 'By the end of 2020 we will produce as much renewable energy as we consume'.[58] This definition, by the way, is a great example of the concept of net use accounting being applied to the goal definition. Rather than claiming 100% renewable energy, IKEA preferred to express its goal in the equalisation of its consumption with renewable energy production.

> If your organisation makes a claim about its use of renewable energy, consider retaining and retiring your RECs. Whilst it costs more, it reduces reputational risk.

Now that you are aware of some of the issues to consider when defining and expressing your goal, in the next chapters we will explore how to analyse the range of renewable energy and energy efficiency options that may be available to you in the pursuit of your 100% target.

Your checklist:

You may be able to delegate or outsource these tasks.

- ☐ Obtain commitment from senior management to a 100% renewable energy goal.
- ☐ Determine what part of your operations will be covered by your 100% renewable energy goal (organisational boundary).
- ☐ Work through your organisational preferences by engaging key organisational stakeholders to establish in broad terms which energy options are acceptable and which not (organisational preferences).
- ☐ Determine what energy forms are part of your boundary (electricity, stationary, and transport energy).
- ☐ Set a target year by which you would like to reach 100% renewable energy.
- ☐ Determine whether you only want one final goal or prefer additional, interim goals.
- ☐ Express your goal.
- ☐ Communicate your goal to your stakeholders.
- ☐ Start thinking about a suitable strategy for your renewable energy certificates (retire vs. sell).

Chapter 3

Making electricity renewable

Now that your organisation has made the commitment to achieve 100% renewable energy and taken some fundamental decisions about the boundaries, timing, and what are and are not acceptable options, let's look at the energy opportunities available.

This chapter details the options to make your *electricity* supply renewable. (Upcoming chapters will explore options for stationary and transport fuels, and discuss battery storage and other innovative opportunities that may become more important in the future.)

The Figure 6 pie chart illustrates how, overall, we currently produce electricity in Australia.

Fossil fuels make up 85% of our electricity production, whereas renewable energy only contributes 15%. Whilst at present less than one-seventh of our electricity comes from renewables, their potential is enormous.

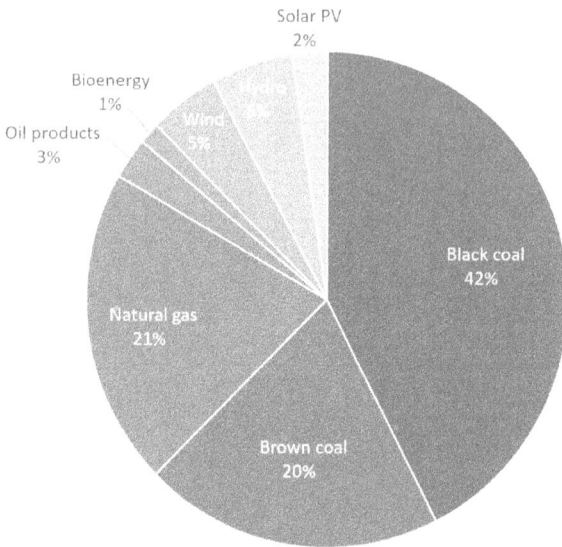

Figure 6: Current electricity production by fuel type in Australia[59]

In its '100 Per Cent Renewables Study – Modelling Outcomes', the Australian Energy Market Operator (AEMO) investigated the maximum recoverable electricity from renewable resources in the National Energy Market (NEM),[60] the biggest electricity network in Australia. AEMO gave consideration to competing land uses, topography, and population density. Only those technologies commercially available now or in the next few years were evaluated. AEMO determined that the current energy demand in the NEM is roughly 200 TWh (terawatt hours) per year. Here is what the study found in regards to the renewable energy generation potential:

- Installing the maximum generation capacity for biomass would cover half of the NEM demand
- Utilising wave power in Australia's oceans would deliver more than what is consumed in the NEM
- Wind farms could produce more than 30 times the energy needed

- Geothermal and Concentrating Solar Thermal (CST) could cover demand almost 200-fold
- Deploying solar PV[61] can potentially produce more than 350 times as much as is consumed.

All up, renewable energy, in both capacity and generation, is potentially about 500 times greater than the forecast National Electricity Market demand. The study also found the operational issues associated with making the grid ready for 100% renewable energy 'appeared to be manageable'.

The following chart graphs the current NEM demand and the potential of renewable energy resources in Australia.

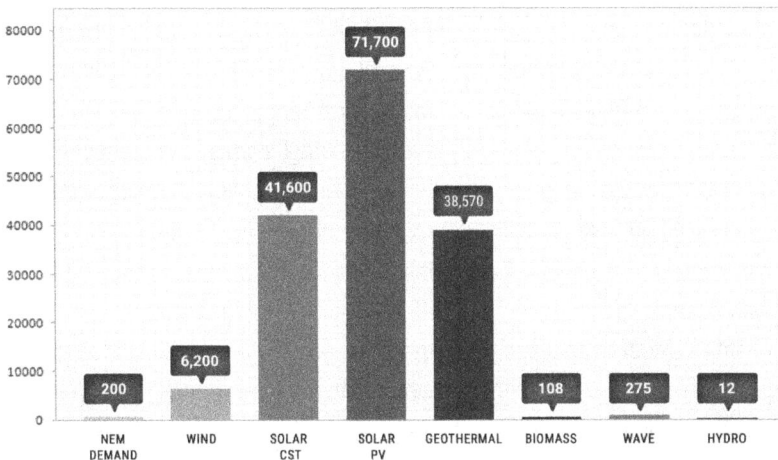

Figure 7: Potential renewable energy generation in Australia in TWh per year

On a country level, there is enormous potential for renewables. But what about on a micro level, for your organisation? Before we delve into the various options that may apply to you, there is an important opportunity to consider first and foremost: energy efficiency, or demand reduction.

Energy efficiency

Organisations moving towards a goal of 100% renewable energy should also maximise their energy efficiency opportunities, especially in light of the Paris Agreement. To limit global warming to well below 2 degrees Celsius, we need to reach zero net emissions from the second half of this century. This means that your buildings and other assets must be 'net zero' by 2050. Whilst renewables will certainly play a role, so will making your facilities extremely energy efficient.

Energy efficiency means to either perform the same activity with less energy input or accomplish more activity with the same amount of energy input. Either way, you achieve more with each unit of energy consumed.

Think of energy efficiency as the cheapest and cleanest fuel you can use, as it is measured and valued as the quantity of energy you do *not* use. The higher the price you pay for your electricity, the greater the value of being more productive with your energy input.

Apart from saving you money, improving energy efficiency means that your renewable energy needs will be smaller, which makes your journey to 100% renewable energy less expensive. It also reduces the environmental impact of manufacturing, transporting, and installing renewables.

I often show my clients the benefits of energy efficiency measures by equating them to the number of solar panels they *do not* have to install. Some organisations also like to see benefits expressed as the number of cars taken off the road.

> Every unit of energy you do not consume does not have to be produced, delivered or paid for.

You can improve energy efficiency by implementing procedural changes, engaging staff, and retrofitting and upgrading equipment. Energy is wasted by leaving appliances and equipment on when not in use, having inadequately controlled temperature or process settings, using old technology, having poor maintenance procedures, or by staff not being aware of the correct operation of equipment. Examples of retrofitting or upgrading equipment include:

- Lighting replacements
- Improving building envelopes to reduce heating and cooling energy demand
- Optimising or upgrading the HVAC system, lighting sensors and timers
- Re-engineering manufacturing processes or implementing new process technology
- Implementing metering and monitoring processes
- Installing variable speed drives on motors used to drive equipment, like fans and pumps.

Even the largest and most sophisticated energy users can find additional opportunities for cost-effective energy savings. To uncover energy efficiency opportunities, you can undertake an energy audit, described in greater detail in Chapter 7.

Focusing on energy efficiency can be a cultural shift for many organisations, and implementing these changes can take time. I recommend implementing an energy management system, like ISO 50001, which works for all organisations, regardless of size, industry, or location, to embed an ongoing culture of energy management and efficiency within your organisation.

Example: We calculated the benefits of energy efficiency improvements across one client's major facilities. Some initiatives were already in progress but we identified additional improvements, like LED lighting, VSD controls, air conditioning optimisation, and heating improvements at one plant. Overall, our energy efficiency improvements saved the client 488 MWh in electricity usage every year. The initiatives equalled the following savings:

- $117,000 annual cost reduction
- 1392 solar panels that did not have to be installed
- Annual energy consumption of 61 households

- 483 tonnes of greenhouse gas emissions
- 141 cars taken off the road.

The costs to realise these savings were estimated to be $580,000, with a cumulative payback period of five years.

Solar photovoltaics (PV) – 'behind-the-meter'

Solar photovoltaic (PV) technology uses energy from the sun to generate electricity. Solar arrays are made up of panels that contain solar cells, which convert sunlight into electricity. The more intense the sunlight, the more power is generated.

The produced electricity is fed into an inverter that converts the low-voltage direct current to an alternating current, which then powers your organisation's equipment. Any excess electricity is sent to the grid through your meter.

The annual solar radiation falling on Australia is about 58,000,000 petajoules (PJ), which is around 10,000 times larger than Australia's annual energy consumption.[62] This makes Australia better suited than most countries to install solar photovoltaics.

In 2014, small-scale solar installations contributed about 15% to the renewable power generation in Australia,[63] and more than 1.6 million households and businesses have installed solar PV. This puts Australia in the lead in terms of the highest proportion of solar PV on roofs of any country in the world.[64]

These figures will continue to increase in the coming years. The adoption of solar PV has enormous growth rates, mostly due to the maturity of the technology, rapidly declining prices, and support by governments at all levels.

According to Bloomberg New Energy Finance (BNEF),[65] the cost of solar power has fallen to 1/150th of its level in the 1970s, whilst the total amount of installed solar has soared 115,000-fold. In its 2016 New Energy Outlook,[66] BNEF further claims that solar energy costs have fallen by 80% since 2008 and will fall another 60% by 2040.

With such rapidly falling prices, it is no surprise that adoption rates of solar continue to soar. Between 2000 and 2015 alone, the amount of electricity produced globally by solar power has doubled seven times over.[67] The following graphic illustrates these growth rates and the cumulative number of small-scale solar, behind-the-meter PV installations in Australia.

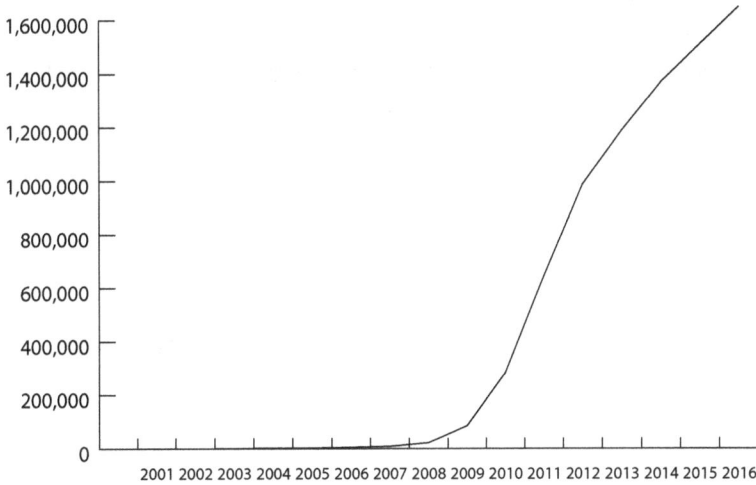

Figure 8: Cumulative number of small-scale solar PV installations in Australia[68]

A 'behind-the-meter' installation means the solar PV system is deployed at the point of the electricity demand (i.e., where people live and work) and not at the side of the grid. Behind-the-meter installations are also known as 'distributed' or 'embedded' generation.

One of their advantages is that power does not need to be transferred over long distances using expensive electrical infrastructure. As a result, behind-the-meter installations displace both the network and retail electricity prices you currently pay. Together, network and retail electricity prices are called a 'bundled' rate.

Every kWh you generate and consume means that you do not have to pay the 'bundled', or per kWh, tariff to your retailer. You may also be able to reduce peak demand charges with a behind-the-meter installation.

This is in contrast to installations in-front-of-the-meter, where your renewable energy project is treated like any other power plant, and where you only get paid (or negotiate) a wholesale electricity price, one much lower than the bundled electricity price. (These systems are described in the next section.)

Below, in Figure 9, is a graphical representation of the two different ways you can deploy your solar system.

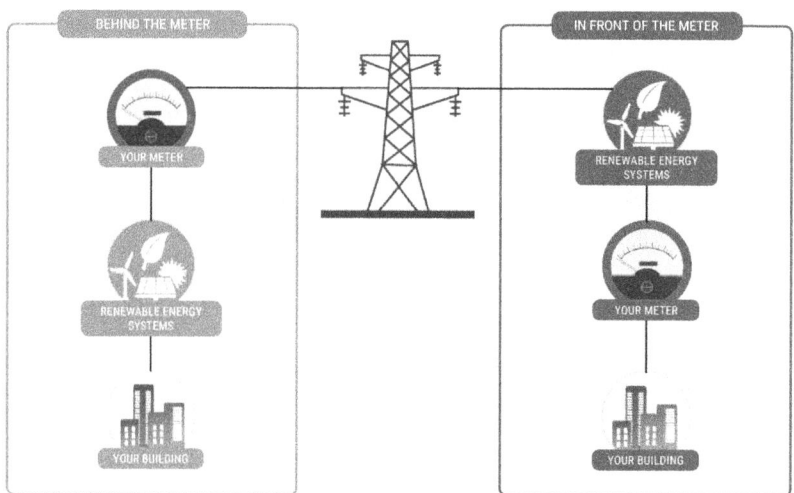

Figure 9: Behind-the-meter versus in-front-of-the-meter installations

Behind-the-meter installations are usually sized so that all the produced energy is consumed on the site, when the sun is shining, but they can also be oversized if you have access to generous feed-in tariffs, or a battery storage system to use the energy at other times.

In Australia, without generous feed-in tariffs, exporting to the grid is currently not cost effective, so there is no incentive to install a system larger than your site needs. This will change once battery storage becomes cheaper, in which case the excess electricity can be fed into the battery, rather than being exported to the grid.

In Australia, the Clean Energy Regulator distinguishes between small-scale (<100 kW) and large-scale generation (>100 kW).[69] No matter whether you install a small or large system, 'behind-the-meter' implementation means that the purpose of the electricity production is to consume it on your site.

PV systems can be mounted on roofs or on the ground; they can also be integrated into a building's façade or roof. Modern PV systems are not restricted to square and flat panel arrays – they can be curved, flexible, and shaped to your building's design.

PV technology in conjunction with storage can also be used for public or street lights. The advantages are that the lights do not have to be connected to the grid, and once you have acquired them, you do not need to pay any electricity costs.

Future improvements for PV technology will be better cellular technologies, greater efficiencies at converting the sun's light to electricity, further declining costs for hardware, engineering, procurement, and construction, and decreasing costs and increased availability of battery storage.

Mid-scale solar PV – 'in-front-of-the-meter'

Mid- or customer-scale solar power plants produce renewable energy for your organisation, which can be used to offset your energy consumption so you can claim 100% renewable energy.

There is no official definition of the size of such a system but it typically ranges between 0.5 MW and 10 MW (as opposed to utility-scale systems, which are much larger).

Mid-scale systems are particularly suitable for organisations with access to underutilised land or water bodies (like the roofs of large parking lots, vacant land, dams, and treated water ponds) near where they need electricity. Mid-scale systems usually connect to the distribution rather than the transmission network, which makes them cheaper to implement than utility-scale plants.

One example of a mid-scale project is the 15 MW Valdora plant[70] on the Sunshine Coast, an artist's impression of which is shown in the following picture.

Figure 10: Sunshine Coast Solar Farm artist's impression,[71] courtesy of Sunshine Coast Council

Other applications for customer-scale solar PV systems are edge-of or off-grid installations for an organisation or town. An example of this is the 1.7 MW (plus optional second stage of 5 MW) solar project at Rio Tinto's Weipa bauxite mine in Australia,[72] which displaces part of its diesel usage to generate power.

The advantages of customer-scale solar projects are low land costs and that they generate renewable energy certificates (RECs). You can decide to sell the RECs to generate income, hold on to them for future use, or retire them to meet your renewable energy target(s). One possible strategy, mentioned in Chapter 2, is to sell the RECs until the time when you intend to reach your 100% renewable energy goal, at which point you start to retire them.

With further improvements in PV technology and continuously falling prices, more organisations are expected to take advantage of this opportunity. Even organisations without access to sufficiently low-cost land can avail themselves of this renewable energy option by entering partnerships with organisations that do. Mid-scale solar projects are also a great opportunity for regional/city partnerships.

Bioenergy

In 2014, bioenergy contributed about 7.6% to Australia's renewable energy generation.[73] It involves the conversion of organic matter, also called 'feedstock' or 'biomass', to energy.

Biomass is produced by plants converting sunlight, water, and carbon dioxide into carbon-containing molecules like sugars, starches, and cellulose. Bioenergy is the chemical energy stored in materials derived from recently living organisms, like plants, animals, and their by-products. Because bioenergy most often uses waste products as input, the technology is also known as 'waste-to-energy'.

Bioenergy accounts for roughly 10% of the world's total primary energy supply. Most of this is consumed in developing countries for cooking and heating, with considerable impact on health via smoke pollution and the environment via deforestation.

Modern bioenergy supply, on the other hand, is comparably small but has been growing steadily in the past decade.[74] Modern bioenergy can provide distributed baseload power and be a great opportunity for industries like agriculture, forestry, food processing, and urban waste management.

Key feedstocks include agricultural wastes, green and food waste, coppicing, or dedicated energy crops. The most common feedstock in Australia is sugar cane waste, but landfill gas, sewage gas, and agricultural and wood waste are also used.[75]

Figure 11: Biomass plant in Eisenstadt, Austria (photography by Claudia Sattler)

Most people consider biomass to be renewable because plants regrow in a matter of a few years, as opposed to fossil fuels that take millions of years to form. When plants grow, they absorb carbon dioxide from the atmosphere. As biomass is burned, the gas is released again, which results in a net effect of zero greenhouse gas emissions.

If whole trees are chopped down for biomass, it could take decades for a replacement forest to grow, which may result in even more greenhouse gas emissions in the short term. However, using mill and pulp plant leftovers to generate bioenergy is an efficient use of woody biomass. The latter are often pelletised to make them easier to handle.

In Australia, the list of eligible renewable biomass energy sources[76] is included in Section 17 of the legislation underpinning the Renewable Energy Target, the Renewable Energy (Electricity) Act 2000. The list includes energy crops, wood waste, agricultural waste, waste from processing agricultural products, food waste, food processing waste, bagasse (crushed sugar cane residue), black liquor (pulpwood waste), biomass-based components of municipal solid waste, landfill gas, sewage gas, and biomass-based components of wastewater.

Only power generated from the biological organic fraction of waste (like green waste or food waste) is considered renewable, whilst the energy generated from fossil fuel-based organics (such as plastics) is not.

There is a large range of bioenergy technologies that convert biomass to energy. The four most common and commercially available conversion processes are: incineration, anaerobic digestion, gasification, and pyrolysis.

Incineration

Here, biomass material is burned to produce heat and/or electricity. Feedstocks are typically wood waste, energy crops or sugar cane. If heat is required at or near the incineration process, the energy plant can be configured as a cogeneration plant. This is done extensively in the Australian sugar industry where bagasse, the sugar cane residue, is used to fuel the sugar mill.[77]

Whenever something is burned, there will be local emissions, such as particulates, which is why environmental laws and regulations usually discourage the burning of wastes for disposal. There are also issues regarding community acceptance of this option. If you are interested in this technology, contact your local Environmental Protection Agency for more information.

Anaerobic digestion

Anaerobic digestion is a combined bioenergy and waste recycling solution and is normally used for feedstocks with a moisture content of 70% or more, such as food waste, animal manure, and wastewater. The solution is ideally suited to farms with animal wastes, abattoirs, food processing, fruit and vegetable packhouses, milk processing, sewage treatment plants, and landfill sites.

The technology involves decomposing organic material without oxygen in a tank (digester) or covered pond. Bacterial processes break down the biomass to form methane, which rises to the top of the tank or pond and is drawn off and combusted to produce heat and/or electricity. In Europe, there is a growing trend towards upgrading this methane gas to biomethane for grid injection (or bioCNG and BioLNG for transport fuel).

The liquid fraction of the digested feedback can be used as a fertiliser, and the solid portion can be used as a soil conditioner.

Pyrolysis

Pyrolysis involves heating biomass in the absence of oxygen to produce oil or gas, and biochar. At temperatures around 500 degrees Celsius, pyrolysis oils are produced that can be used as a boiler fuel.

At higher temperatures of 700 to 800 degrees Celsius, more gas is produced, which is drawn off and combusted to generate heat and/or electricity. It has a calorific value of about half that of natural gas and may be used to fuel engines and gas turbines without modification.[78]

Biochar is a stable form of carbon that can be used to stabilise and condition soil. Biochar can store carbon in the ground for long periods, so the process can remove carbon dioxide from the atmosphere.

Advanced gasification

Thermal gasification is normally used for low moisture feedstocks, such as woody biomass and crop residues. If feedstocks contain much water, they need to be dried beforehand.

The gasification of organic waste produces gas and biochar, which, again, can be used as a fertiliser replacement. The low calorific value of the gas requires using greater volumes to achieve the same energy output compared with using natural gas.

Gasification is also used for municipal waste that has been stripped of recyclable materials or hazardous waste. The waste is gasified in a high-temperature environment, creating synthesis gas, or syngas, which can be used to generate electricity and heat.

Wind

Windmills have been used for hundreds of years. In their current form, wind turbines have been available since the 1980s, but the technology took off around the year 2000, when cost reductions began to make it a viable generation source for utilities.

Wind power is currently one of the cheapest sources of large-scale renewable energy.[79] In 2014, it contributed 31% to the renewable power generation in Australia,[80] a number destined to grow in the coming years with further price falls and technology improvements. Worldwide, the amount of wind-based electricity produced globally has doubled four times over just between 2000 and 2015, according to Bloomberg New Energy Finance.[81]

Wind turbines are designed to convert the energy of wind movement into electricity. The most efficient wind turbines are those with three blades. As wind blows at a 90-degree angle to the blade, areas of lower and higher pressure are created above and below the blades, causing them to rotate, which drives a generator behind the blades. These are called 'horizontal axis' wind turbines which, according to the laws of physics, are more efficient than 'vertical axis' wind turbines.

Some people worry about the health impact of wind turbines but there are innumerable studies that refute this misconception. The Australian Medical Association, for instance, released a statement on Wind Farms and Health,[82] in which it expresses their its view that there are no adverse health effects on populations residing in the vicinity of wind farms.

The windiest areas are typically coastal regions at mid to high latitudes, and mountainous regions. In Australia, we have some of the best wind resources in the world, primarily located in western, south-western, southern, and south-eastern coastal areas, but also extending hundreds of kilometres inland, and including highland regions in south-eastern Australia.[83]

Figure 12: Wind turbine with viewing platform in Bruck an der Leitha, Austria

New developments with large-scale wind farms are increasing average hub heights. The higher the centre, the stronger the wind is. It is predicted that new hub heights could be as much as 170 metres, which means the blade tip would extend as high as 250 metres. For comparison, the height of the Eiffel Tower is 300 metres.

The wind at these heights comes more regularly from the same direction, which means less turbulence and more power production. The larger blades also spin more slowly, making it easier for birds to see and avoid them.

Another development is offshore wind farms, where the turbines are anchored in the ocean. This technology is used widely in England and Germany. Although currently more costly than onshore wind turbines, offshore wind turbines have the advantage of avoiding the noise and visual impacts of onshore developments, and the wind tends to be stronger and more continuous. In Australia, we have the second-largest offshore wind energy resource in the world, second only to the Russian Federation.[84]

Wind farms are usually deployed at a utility scale and you can purchase the power from these farms through PPAs (power purchase agreements – you can read more about these agreements in Chapter 10), or through your retailer. However, you can also deploy wind at a small scale, in the form of micro turbines. Most small-scale turbines are used in remote regions where the grid is unavailable, but they can also be integrated into buildings.

For micro developments on buildings to perform optimally, the height needs to be more than 100 metres (or 30 storeys), as winds get stronger with increasing height and there are fewer obstructions. Even so, these developments come at a high cost and generate comparatively little energy, which makes urban wind less viable than large wind farms in regional areas.

Hydro

The term 'hydro' comes from the Greek word *hydra*, meaning water. Hydroelectricity (often abbreviated to 'hydro') is a well-developed renewable energy technology that has been around for more than a century.

It is the world's primary source of clean energy and provides just over 16% of global electricity production.[85] In 2014, hydro contributed about 46% to renewable power generation in Australia,[86] the largest proportion of any of its current renewable energy technologies.

Hydro uses flowing water to spin a turbine connected to a generator that produces electricity. The amount of power produced depends on the volume, and the height (called 'head'), of the water above the turbine.

In times of drought, water to hydroelectricity systems is limited, and electricity output falls, such as occurred in Tasmania's hydro-power system in early 2016,[87] or in New Zealand, in 2008.[88] However, under normal (no drought) circumstances, hydro boasts a significant advantage in that it provides dispatchable energy,[89] as hydro generators can start up and supply maximum power within 90 seconds.

Hydroelectric systems are classified according to the size of their generating capacity. There is some variation in the definition of 'large' but most plants over 10 MW are categorised as large. Depending on local availability, you may be able to purchase renewable power from large hydro plants through PPAs or through your retailer. Small systems are further divided into 'mini' and 'micro'.

The main generating methods are 'conventional', pumped storage (which is typically not fully renewable due to the pumping that is involved) and run-of-river. Conventional hydroelectric power stations need dams to store the water required to produce electricity. These dams are often built to hold irrigation or drinking water, and the power plant is included in the project to ensure maximum value is extracted from the water. However, a disadvantage of constructing large dams is that people living in the area to be flooded are displaced, as are the local fauna and flora.

Run-of-river hydropower schemes have little or no reservoir capacity, and rely on water upstream, with any oversupply passing unused.

Pumped storage hydro uses two reservoirs at different elevations. During off-peak times, water is pumped from the lower to the higher reservoir. During peak demand, the water is released back into the lower reservoir through a turbine, which generates electricity. Pumped-storage schemes provide an important means of large-scale energy storage.

Small-scale hydropower systems can use dams but most rely on naturally flowing water, such as rivers, where part of the flow is diverted for power generation. Because they are small, they have fewer negative environmental impacts compared to large-scale plants. Small-scale systems are often used as stand-alone systems and are not connected to the main electricity grid. Austria is home to many of these small-scale systems.

Small hydro systems can also be installed in water and treated wastewater pipes. An example of this is the North Head Sewage Treatment plant, run by Sydney Water. The unit harvests power from treated wastewater falling down a 60-metre shaft.[90]

Marine energy

With oceans covering roughly 75% of the world's surface, such abundance makes marine energy an increasingly attractive form of renewables. Compared to sun and wind energy, marine power provides a predictable, high-density source of renewable energy.

In addition to its enormous availability and high predictability, it is located close to many major cities and has little visual impact. Companies like Carnegie Wave Energy[91] have made great progress developing this technology. However, marine energy currently plays only a small part in the renewable energy generation mix in Australia.

There are two principal forms of marine energy: tidal energy and wave energy. Tidal power captures the energy of the gravity-driven tides and generates electricity using the regular local flows of the tidal cycle. The Kimberley and Pilbara coasts of northern Western Australia see the largest tides in Australia. Other potential sources of tide power are the Torres Strait off the coast of Darwin, Broad Sound in Queensland, and Bass Strait in Tasmania.[92]

Wave power plants harvest the energy in the up and down motion of waves driven by the wind and convert it into electricity. Wave energy is strongest where there are trade winds and ocean swells. In Australia, our greatest wave energy resources are along the southern coastline.[93]

Depending on local availability you may be able to purchase renewable marine energy from your retailer, or through PPAs.

Geothermal

Despite its great potential, geothermal energy currently plays only a small part in the renewable power generation mix in Australia. Geothermal technology uses the Earth's natural internal heat to generate electricity (and heating). For the moment, we will focus on electricity generation. The next chapter covers the use of geothermal heat pumps for heating or cooling needs.

Geothermal energy may be stored in granite rocks, which are often called 'hot rocks' or 'enhanced geothermal'. Wells are drilled to a depth of three to five kilometres below ground and water is pumped into the wells, where it is heated to a temperature of up to 300 degrees Celsius. The resulting steam is pushed to the surface, where it drives a turbine to produce electricity.

However, the most common form of geothermal electricity globally is 'low temperature' geothermal. With this technology, heated water is brought to the surface from hydrothermal reservoirs, using boreholes, and cooled water is pumped back into the ground to maintain the water table and pressure. This technology may be of interest to industries and companies operating in remote off-grid locations with access to a sedimentary aquifer.

Some big advantages of geothermal power are that it provides dispatchable power and can be switched on and off, based on demand. Countries located above geothermal hot spots can generate significant amounts of electricity from this energy source.

Iceland, for instance, obtains 25% of its total electricity generation from geothermal sources. In the Philippines and Kenya, geothermal energy represents around 17% of power production. In New Zealand, geothermal energy provides 13% of the country's electricity demand, and has been described as New Zealand's most reliable renewable energy source, above wind, solar, and even hydroelectricity, due to its independence from the weather.[94]

In Australia, the only geothermal plant currently in operation is based in Birdsville, Queensland.[95] The 80 kW system draws water at 98 degrees Celsius, up a 1.28-kilometre deep bore from the Great Artesian Basin, and produces electricity through an organic Rankine cycle.[96]

Depending on local availability, you may be able to purchase renewable geothermal energy from your retailer or through PPAs.

Purchasing renewable energy

There are several reasons you may need to purchase renewable energy. You may rent your business premises and have limited opportunities for onsite renewables. Your embedded generation may not meet all of your site's demand.

You may need a bridge option to enable a quick solution to meet renewable energy targets whilst you implement longer-term solutions. It may just be too costly or uneconomic to develop renewables onsite, and partnering with developers or organisations with access to less costly land or better resources might be more attractive.

Alternatively, you may just prefer to buy renewable energy rather than install energy solutions. Even though you might not add renewable energy capacity to your site, buying green power still helps create more certainty in the marketplace, helps finance the building of renewable energy assets, and encourages others to invest.

In most regions, energy retailers offer plans to purchase renewable, ('green') energy. Some might give you the option to specify the percentage of renewable as opposed to the 'black' energy you buy. Others might sell you a particular number of kWh from renewable sources. Usually, this attracts a premium to grid-supplied electricity.

As an alternative to purchasing green power from your retailer, you can also directly buy renewable energy certificates/RECs in the spot (short-term energy) market.[97] Many businesses take extra precautions to make sure the RECs they purchase comply with stringent environmental and social principles. Some organisations will only accept local renewable energy projects; others may want to ensure the renewables they purchase are from new projects, rather than existing ones.

SAP, a German software corporation, for instance, requires its power plants be no more than 10 years old and will only consider renewable electricity from biomass that is disconnected from coal or other fossil power plants and if the biomass itself is not related to deforestation.[98] Steelcase, a producer of office furniture, only buys RECs generated by projects with zero carbon emissions.[99]

Another option to purchase renewable energy is to enter into a power purchase agreement (PPA – see more on these in Chapter 10). There are two types of PPAs: 'Unbundled PPAs' and 'bundled power and RECs' agreements. Unbundled PPAs allow an organisation to receive the benefits of renewable energy without receiving the energy itself. Under this agreement, you purchase the RECs from renewable energy developments, whilst maintaining the same contractual arrangements with your retailer.

Bundled power and RECs are PPAs where you purchase both the RECs and the energy. You enter into an agreement with a renewable energy developer, potentially in conjunction with your energy retailer, and pay your retailer a contracted price for an agreed volume of power, as well as the RECs that come with it.

Some organisations choose bundled power and RECs agreements because they are concerned about the 'additionality'[100] of their purchase. Unbundled RECs by themselves are sometimes seen as not providing enough impetus for further growth in renewables.

The disadvantage of the bundled approach is that if your organisation's demand is higher than the output from the renewable energy project, you must purchase the remaining RECs on the open market, exposing you to price fluctuations. Conversely, if your demand is lower than the output, you can sell the RECs and generate income.

The food company Mars, for instance, partnered with a developer on a 200-MW, 118-turbine wind farm in Texas that came online in 2015. The installation produces electricity equivalent to 100% of the company's US power needs.[101]

In Australia, the ACT Government is committed to 100% renewable energy by 2020. It has used reverse auctions to build large wind farms in Victoria and South Australia, which sell their electricity output back to the Government via PPAs.

At a conference I presented at, the ACT's Minister for the Environment and Climate Change was asked why the Government sources its renewable energy from other states. The answer was that, previously, the ACT got all its black power from the Hunter Valley in NSW, or the Latrobe Valley in Victoria, since there are no big conventional power plants in the ACT. Sourcing energy from other states was thus not different to the previous model, except that the energy was now renewable.

Depending on where you are located, you may be able to access aggregated buyer groups. One example is WWF, which coordinates a

Renewable Energy Buyers Forum to enable organisations to more easily purchase renewable energy.[102]

It depends whether you would rather have the renewable energy development close to your energy demand or are more interested in getting the best financial outcome for your investment.

All in all, there are a number of options available to you to make your electricity consumption renewable. First and foremost is to reduce your energy consumption, after which you can consider producing renewable power by installing solar, small-scale wind, hydro, geothermal, or biomass systems. You can also purchase renewable energy from sources outside of your immediate reach, through PPAs, your retailer, or by buying RECs. Chapter 11 discusses how to best use these opportunities to reach 100% renewable energy.

Your checklist:

You may be able to delegate or outsource these tasks.

Examine, at a high level:

☐ Your electricity-consuming equipment and how energy efficiency might apply

☐ How behind-the-meter and in-front-of-the-meter installations would apply to your operations

☐ Your electricity-consuming equipment and the readiness and availability of renewable energy options

☐ Whether you will need to purchase renewable energy as part of your pathway to 100% renewable energy.

Chapter 4

Making stationary (heating) fuels renewable

The previous chapter explored options to make your electricity supply renewable. The focus of this chapter will be to make renewable stationary *fuels*, ones used to produce heat. Stationary fuels are typically used for equipment that does not move or have a licence plate – like boilers. Stationary fuels, such as natural gas, liquefied petroleum gas (LPG), and coal, are used by industry for process heat ranging from below 100 to more than 1000 degrees Celsius, and also as a chemical feedstock.

Whilst the biggest users of stationary fuels are the metals, chemical processing, wood and paper, and ceramic industries, where many processes require high temperatures, this chapter focuses on those organisations requiring heat below 150 degrees Celsius, like food and beverage manufacturing, or the commercial sector. This is because renewable energy technologies are most suitable and available for lower temperature needs.

Many manufacturing processes use heat, such as for cleaning, pasteurisation, drying, various cooking techniques, space heating, and domestic hot water. The commercial sector, which includes hospitals, offices, hotels, schools, and shopping centres, typically use natural gas for space heating and hot water.

There are four principal options for making heat renewable. A necessary first step before introducing renewables is to reduce the amount of energy used to generate heat, or decrease the heat required. Being more energy efficient will help you understand energy use and maximise the benefits from your renewable energy investment. You can then either use energy from the sun, electrify your equipment and use heat pumps, utilise geothermal heat directly, or switch to biomass as a fuel. Where stationary fuel sources cannot readily be transitioned to renewables, you can purchase carbon offsets in the interim.

Some of the options in the following sections are based on findings in the Greenpeace report 'Energy [R]evolution'[103]and IT Power's report for ARENA on 'Renewable Energy Options for Australian Industrial Gas Users',[104] both from September 2015.

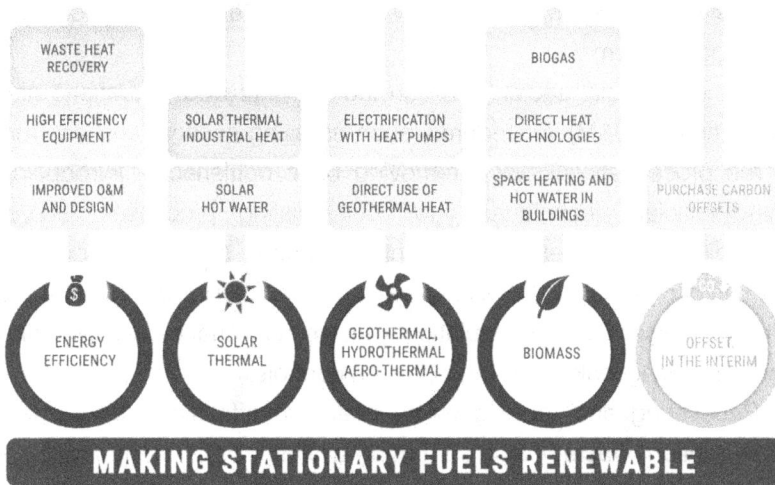

Figure 13: Options for making stationary fuels renewable

The following sections describe the currently available technologies that can help make your stationary fuels renewable. These opportunities will become increasingly attractive as the cost of renewables comes down even further.

When you read through these options, please bear in mind there is no one-size-fits-all solution and that it is best to investigate the suitability of your opportunities in detail. A qualified energy consultant can help you with this task.

Energy efficiency

As with the previous chapter, before you switch fuel sources to make your energy supply renewable, you should first consider reducing your energy demand. Whilst this does not qualify as a renewable option per se, it is an important first step before you consider renewables.

The following options to improve your energy productivity with respect to stationary fuels are a good starting point, but they are not exhaustive, and you may find many more opportunities once you start investigating them.

In the industrial sector, you can look at improved operations and maintenance (O&M) procedures, such as using sensors and controls to run processes efficiently, removing plant bottlenecks, introducing energy-efficient techniques into your existing processes, and/or upgrading to more energy efficient equipment, such as modulating burners or condensing boilers, for example. You can reduce heat losses from processes through insulation, regular maintenance, by detecting and eliminating leaks, and by improving controls.

Recovering and reusing waste heat can effectively reduce fuel consumption. Waste heat is commonly available when stationary fuels are combusted, such as in boilers or ovens. Waste heat may also be recoverable from cooling processes, such as from chillers and cooling towers.

Heat from these processes is normally released into the environment but can be used to reduce the energy consumption of the process itself, or to provide heat for other processes or systems such as space heating.

For boiler and steam systems you can also recover or avoid waste heat through effective maintenance of steam traps, maximising condensate return, or use of blowdown recovery systems.

In the commercial sector, the demand for heating can be reduced by looking at improvements in the building design, or through re-commissioning, including opportunities like heat loss reduction, foundations, building orientation, shading, insulation, glazing, and solar heat gain.

You can also consider implementing or re-commissioning your building management system (BMS), which can automatically regulate temperature control and other systems in your buildings. In regards to the hot water consumption, you can achieve a substantial reduction in energy use by implementing flow control fixtures on taps and showerheads.

Solar thermal technologies

Solar thermal collecting systems are based on the fact that the sun heats up fluid contained in a dark vessel. Solar thermal technologies can provide energy for a wide range of applications in domestic and commercial buildings, swimming pools, for industrial process heat, in cooling, and desalination for drinking water.

Solar hot water

Solar thermal technologies convert sunlight into heat or thermal energy. Solar hot water uses a flat plate or an evacuated tube collector to harness solar energy into heat. The solar collectors are linked to an insulated storage tank.

Solar hot water typically needs a booster element to make sure there is hot water when the solar radiation is low, or in cold areas. To make this solution fully renewable, the booster energy needs to come from renewable sources, like solar PV.

Solar hot water is typically used for domestic purposes, but it can also supply the hot water for commercial and industrial premises to meet domestic hot water, cleaning, and low-temperature process needs. It is commonly used for heating swimming pools, through unglazed, glazed, or evacuated tube collectors.

On my quest to make my personal energy needs fully renewable, I had a solar hot water system installed on the roof of my house. Because the solar PV panels took up considerable space, I settled on evacuated tubes, which are more efficient than flat panel systems (but also more expensive).

The following chart shows the adoption of small-scale solar hot water in Australia, which includes air-source heat pump installations, making up about 20% of the total.

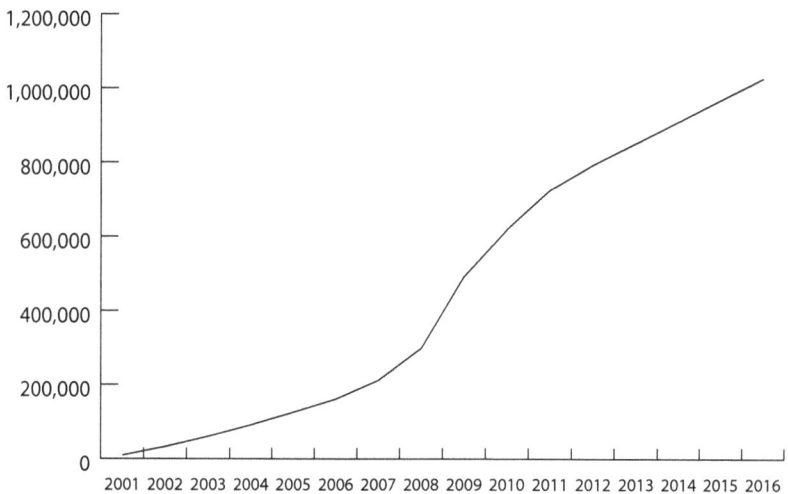

Figure 14: Cumulative number of small-scale solar water heaters and air-source heat pump installations in Australia[105]

Solar thermal industrial heat

There are many organisations, especially in the manufacturing sector, with a large demand for heat for their processes. Depending on the temperature level needed, different collectors have been developed to serve the requirements for process heat.

Lower temperatures are available from unglazed, glazed, or evacuated tube collectors. More complex and costly systems are needed for higher temperatures, like parabolic troughs and Fresnel reflectors, which can deliver temperatures between 60 and 450 degrees Celsius.

One case where a solar thermal plant was implemented is the Green Brewery, in Austria. Occupying an area of around 1400 square metres, the solar thermal plant in Leoben generates almost 30% of the thermal energy required for the mashing process. A 200-cubic-metre energy storage tank allows excess thermal energy to be collected and saved for future use, ensuring a continued energy supply to operate the brew house.[106]

Figure 15: The solar thermal plant at the Green Brewery in Leoben, Austria (photo usage courtesy the Green Brewery)[107]

Another example is the De Bortoli winery in Australia. As part of a bottling line expansion, the De Bortoli winery installed a large solar thermal evacuated tube collector (about 200 kW$_{th}$) at its Griffith winery in 2013.[108] As a result, De Bortoli can produce 12,000 litres of hot water at 90 degrees Celsius for 80% of the year.

The evacuated tube collectors are mounted at a tilt angle of 37 degrees to optimise their performance in high demand periods, and two 6,000-litre stainless steel storage tanks contain the hot water.

Electrification and geothermal heat

Direct use of geothermal energy

(Deep) geothermal heat from aquifers or reservoirs can be used directly in hydrothermal heating plants to supply heat demand close to the plant or in a district heating network for several different types of heat.

Heated water from geothermal wells is most commonly used for commercial and industrial heating and cooling where geothermal resources are available. Geothermal wells are drilled to a depth of several hundred metres, where temperatures are sufficient to heat water to desired temperatures.

One of the largest geothermal heat users in the world is the Kawerau timber processing plant, in New Zealand. The geothermal field has provided steam since 1957 and is the cheapest source of energy in Kawerau. The steam is used by several power stations to generate electricity, but the heat is also directly used by various companies for boiler feedwater heating, timber drying kilns, and paper drying.[109]

Electrification and heat pump usage

For some applications, you can evaluate whether it will make sense financially to change your existing gas-fired heating system to electric heat pump technology powered by renewable energy. A heat pump is a device that transfers heat from one fluid at a lower temperature to another at a higher temperature. To make the technology fully renewable, the electricity has to come from renewable sources.

Heat pumps have become increasingly important in buildings and heated swimming pools but can also be used for industrial process heat. Industrial heat pumps use waste process heat as the heat source and deliver heat at a higher temperature for use in industrial processes, heating or preheating, or for space heating and cooling in industry.

Heat pumps are distinguished by the heat source they use. Ground-source heat pumps exploit the energy stored in the ground at depths from around hundred metres up to the surface. Ground source heat pumps can be used for residential and small commercial buildings.

They are rare in Australia, and therefore expensive, but one example where a ground source heat pump was installed is Lithgow Hospital.[110] When the hospital evaluated its options, the geothermal system had the highest capital but the lowest running cost.

Water source heat pumps are coupled to a (relatively warm) water reservoir of around 10 degrees Celsius, such as wells, ponds, rivers, or the sea. Examples of an open-loop aquifer system can be found at the Sydney Opera House, the Australian National Maritime Museum, and Barangaroo South[111] which use the water temperature in the Sydney Harbour for their air conditioning systems.

These types of heat pumps do not have widespread application, as opposed to air-sourced heat pumps. The latter use the outside air as a heat source and are mostly used in air conditioning equipment.

The efficiency of a heat pump is described by the Coefficient of Performance (COP), the ratio between the useful heat output and the energy consumption of the unit. The higher the COP, the more energy efficient the equipment is.

Let's use an example where heat pump technology is used to air condition a building, where reversible heat pumps can be operated both in heating and cooling modes. If the COP is four, it means that for every single kWh put into the heating system, 4 kWh of heat are produced.

If an organisation has a big, suitable roof space, then the installation of a solar PV array can provide the renewable electricity for the running of the equipment. This can be combined with batteries to meet the heating and cooling demands during early mornings and late afternoons, when solar output is low.

Most heat pumps use 410A as a refrigerant, which is a potent greenhouse gas. Investigate whether you can get heat pumps that use refrigerants with a lower global warming potential.

Biomass heating technologies

Biomass can provide a large temperature range of heat and be transported over long distances, which is an advantage compared to solar thermal or geothermal heat. On the other hand, the further the biomass has to be transported, the more unsustainable and more expensive the solution will be. The burning of biomass also produces local particulate emissions.

Such considerations have led to biomass heating being mostly deployed as decentralised systems for space heat and hot water, or as cogeneration systems for industry and district heating.

IKEA Industry Group, for instance, supplies 82% of the heat energy it requires from biomass.[112] Nestlé Australia uses waste products like coffee grounds and sawdust in a boiler in its Nescafé factory in Gympie. The biomass facility covers 70% of the site's energy needs.[113]

Space heating and hot water in buildings

For colder regions, firewood and pellet boilers can provide space heat and hot water. 'Pelletisation' has recently become widespread in North America and Europe, whereas Australia does not yet have an established supply chain.[114]

Woody biomass (which includes wood chips and crop waste) is pressed and extruded into pellets, which have a higher energy density and lower moisture content, making transport and storage more economical and efficient. Small-scale pelletisation plants can be set up in large agricultural areas or on industrial sites, like food manufacturers that use grains.

I personally experienced the transformation from fossil fuels to renewables for space heating and hot water needs when I grew up in Austria. In the early 1970s, we had a big pile of coal in our cellar to heat our house with.

In the late 1970s, we replaced the coal with a plant room that held our boiler, run by oil. I remember how, before every winter, a truck would deliver the oil via a long hose inserted through a tiny basement window.

Recently, that plant room in my parents' house made way for a sauna, and the central heating and hot water is now provided from a biomass plant in the village's centre, fed by wood waste from the local area.

Direct heating technologies

There are three main technologies for direct heating: cogeneration, direct combustion, and 'upgraded biomass', like pyrolysis and gasification.

Direct combustion of biomass can be used for temperatures between 80 and 800 degrees Celsius, whereas biomass gasification and combustion can be used from 80 to morew than 1000 degrees Celsius.

Cogeneration increases the benefit of using biomass if the waste heat can be used efficiently. Combustion systems are primarily used for steam and hot water production. Gasification or pyrolysis involves heating biomass via partial oxidation to high temperatures so that most is converted to gases or oils. These can be combusted for process heat, used in engines, or, after further investment, converted to chemical feedstocks.

Biogas

As described in the previous chapter, under 'bioenergy', biogas can be recovered from wastes that are not otherwise recycled, such as residential and commercial waste, sewage, landfill sites, livestock manure, and agricultural and forestry waste products. The resulting biogas can be used as a natural gas replacement for heat applications from 80 to more than 1000 degrees Celsius. Other names for biogas are 'syngas', 'synthetic natural gas', 'green gas', 'renewable gas', or 'biomethane'.

In some countries, you can purchase biogas from your retailer. One example is Green Energy UK, which supplies customers with biogas from anaerobic digestion plants across the UK using biodegradable matter, such as food waste, sewage, energy crops, and residues from food manufacture.

Figure 16: The biogas plant at the Green Brewery in Leoben, Austria photo (photo usage courtesy the Green Brewery)[115]

Offset your stationary fuel consumption

The goal of achieving 100% renewable energy is challenging, but achievable. However, as you embark on your journey, some plans may be delayed, your energy projects may fall short of your desired targets, organisational circumstances might change, or your current technology might not be easily transitioned to renewables. For instance, renewable energy options may not be as readily available or attractive for process heating at high temperatures (i.e., more than 300 degrees Celsius).

Purchase carbon offsets

Carbon offsets may be a good way to ensure targets are met whilst maintaining the pathway to 100% renewables. To make use of this opportunity, you would match the greenhouse gas emissions resulting from your stationary fuel consumption with the purchase (and retirement) of carbon offsets.

When you purchase carbon offsets, consider buying offsets that meet stringent standards to make sure the underlying projects are real, achieve a permanent, verifiable reduction in carbon emissions, and are in addition to business-as-usual operations. Two examples of quality carbon offset standards are the Voluntary Carbon Standard[116] and the Gold Standard.[117]

Ensure the underlying carbon offset projects fit in with your corporate policy and pick a suitable project location (in your country or overseas). You can consider buying carbon offsets based on renewable energy projects, since your goal is to make your stationary fuel consumption renewable.

You can buy carbon offsets through your preferred broker, become a member of a carbon exchange and buy directly, or go to tender for your carbon offset purchase. Select a reputable carbon offset provider. Ask what carbon offset registry they use and enquire about their retirement procedure.

Your checklist:

The following list only applies if stationary energy is part of your boundary. You may be able to delegate or outsource these tasks.

Examine, on a high level:

☐ Your processes and how energy efficiency could apply
☐ Your processes and the readiness and availability of renewable energy options
☐ Whether you need to buy carbon offsets to achieve 100% renewable energy for your stationary energy needs.

Chapter 5

Making transport fuels renewable

According to the International Energy Agency (IEA), the direct combustion of fossil fuels for transportation purposes accounts for 19% of global primary energy use and 23% of energy-related carbon emissions.[118] Transport energy demand has increased at an annual rate of just under 2% over the past decade[119] and will keep rising as our world's population and car ownership rates grow. In Australia, there is a continued reliance on road transport for freight and supply chain logistics.

Just like burning fossil fuels to generate electricity or heat, combusting petroleum, diesel, and LPG contributes to climate change and air pollution. Air pollution leads to smog and particle pollution, which has been linked in some instances to cancer, as well as damage to our immune, neurological, reproductive, and respiratory systems.[120] Along with harming human health, air pollution also causes acid rain, eutrophication, haze, ground-level ozone, and affects the health of animals.

While the transition to renewables is well under way for electricity consumption, transport energy lags behind and has the lowest renewables share among all sectors. Renewable energy represents about 3% of the total final energy consumption in transport. About 2.8%

comes from liquid biofuels, and 0.2% from renewable energy through electric mobility.[121]

Your organisation may have a light vehicle fleet, consisting of cars and maybe some vans and utility vehicles. You might also have a heavy fleet, consisting of trucks or plant assets, which most likely run on diesel.

To reduce the emissions from your vehicles, a combination of energy efficiency and renewable fuels is required. As shown in the diagram below, the opportunities for making transport energy renewable lie in reducing your fuel consumption, changing your fuel supply, and in purchasing carbon offsets in the interim, until renewable fuel supplies are more readily available and at the quality needed.

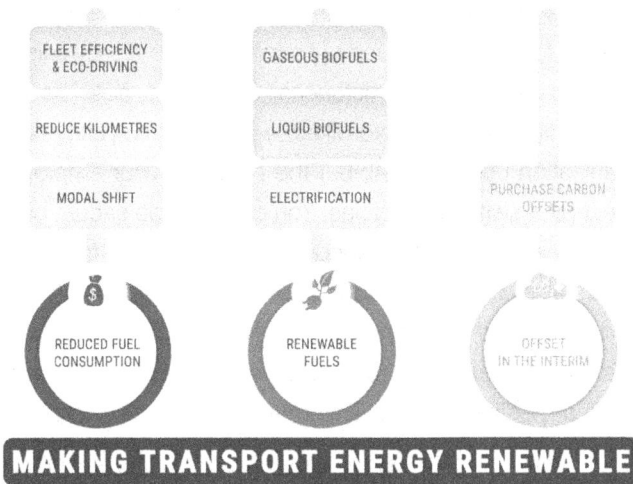

Figure 17: Options for making transport energy renewable

Reduced fuel consumption

Before you investigate changing your fleet to biofuels or electric-powered vehicles, examine opportunities to reduce fuel consumption first. This will save you money and may mean you can change fewer vehicles over to renewable energy.

Modal shift

A modal shift means switching from a high-energy-intensive transport method to a lower-energy-intensive one. Passenger vehicles, for instance, move a person with 80% more energy than a bus; rail moves a tonne-km of freight with 90% less energy than road freight.[122]

What this means for your organisation is that rather than fly, your employees may be able to take the train, or teleconference. Rather than drive their cars, they may be able to drive a motorbike or bicycle, a practice some couriers have adopted.

Instead of having employees (or your customers) drive to your office or facility, consider providing a shuttle bus service from public transportation hubs. Rather than sending physical documents, use cloud services instead. One of the most relevant and substantial shifts for industry is to have trucks use rail services, where they exist.

Depending on your circumstances, it may be possible to change some of your light fleet to motorbikes, e-bikes, or bicycles. Not only will your organisation use less fuel, but the health of your employees might also benefit by a switch to bicycles. You can provide secure bike storage and shower facilities at work to encourage more people to make the shift.

You can also incentivise your staff to use public transport where possible by issuing them travel passes, encouraging them to walk short distances rather than drive, and use video conferencing rather than travelling to a meeting. Simply raising awareness is a good start.

Reduction in kilometres travelled

If you can reduce the number of kilometres your fleet travels, you can reduce fuel consumption (and thus your costs). This can be achieved in numerous ways:

- Provide a fleet that is shared
- Evaluate your logistics and potentially consolidate deliveries
- Arrange two-way loads to reduce the number of times your fleet has an empty return trip; if your organisation does not have any return loads, maybe you can partner with nearby businesses that do, and share your trips.

Behaviour change is important, too. Encourage your staff to re-think whether they need to undertake a trip if other options are available. Conducting annual staff surveys, or reporting on organisational trends or your progress towards 100% renewable energy, helps to increase the awareness of your goal. You may also allow your employees to work from home some or all of the time to avoid the commute to your office.

If you relocate or open a new office, you can map out all public transportation and non-vehicle transport routes to encourage other modes of transport. In the future, self-driving or autonomous cars may also help to reduce the kilometres driven through improved route planning, avoiding congested traffic areas, and vehicle-to-vehicle communication.

If your business regularly delivers goods, you can better utilise the cargo space to transport more in one trip. Maybe options exist to reduce the packaging or you can optimise truck and container loads to maximise fill rates. When IKEA develops new products, product designers consider packaging solutions right from the start, to increase filling rates.[123]

It is important to extend your efforts to outside your organisation and engage with your suppliers to see how you can reduce fuel consumption in your supply chain. Lots of organisations, like H&M, now engage their logistics supply chain to 'green' their transport.[124]

Efficiency improvements

Fuel efficiency is measured by the amount of fuel it takes to move a passenger or a tonne of freight over a certain distance. For light vehicles, this is often expressed as the litres of fuel per 100 kilometres of travel. For heavy fleet, this can be expressed in litres per kilometre, or in litres per tonne.km. You can also use greenhouse gas emissions, for instance kilograms (kg) of carbon emissions per kilometre or tonne.km travelled.

Overall, vehicle efficiency is a combination of engine efficiency and other factors, such as transmission efficiency, vehicle weight and aerodynamics (reduced drag), driver behaviour, and driving conditions.

There is an ongoing move to make vehicles more efficient, especially in government jurisdictions with stringent vehicle fuel economy or tailpipe carbon emissions standards,[125] like in the US, Europe, Japan, China, Canada, South Korea, Mexico, Brazil, and India. This mounting regulatory pressure is forcing car manufacturers to invest heavily in low-emission technologies.

Increasing efficiency standards does not just relate to light duty vehicles, but also to heavy-duty vehicles, like trucks. The US, Japan, and Korea all implement efficiency-based standards for heavy-duty vehicles.

How can you improve the efficiency of your fleet?

In your purchasing policy, favour smaller, lighter, and more efficient cars or trucks as much as possible. A great website to determine the fuel efficiency of a particular vehicle is the GreenVehicleGuide.[126]

Traditionally, vehicle fleets in Australia were dominated by six-cylinder cars, due to fringe benefit taxes. However, most clients I work with have already switched from six- to four-cylinder cars, and some have purchased hybrid cars.

'Standard' hybrid cars have both an internal combustion engine and an electric motor. Hybrid cars use regenerative braking, which recaptures the energy normally lost during braking. The electric motor provides power to assist the engine to accelerate, overtake other vehicles, or climb hills, which results in greater efficiencies.

In some hybrids, the electric motor alone propels the vehicle at low speeds, where internal combustion engines are least efficient. Hybrid cars are most energy-saving for city driving, whereas for highway driving, there is less braking, making efficiency less substantial, as the combustion engine has to operate longer.

There are also 'plug-in' hybrid vehicles (PHEVs) which recharge the battery through both regenerative braking and plugging into a source of electrical power. PHEV examples are the Ford C-MAX Energi, the Chevy Volt, Mitsubishi Outlander, and the Toyota Prius Plug-in.

Apart from changing to more efficient vehicles, other tips include not to unnecessarily transport heavy goods in your vehicles, and to make sure vehicles are regularly maintained. A well-tuned engine can save 10% on fuel consumption.[127]

Tyres should be properly inflated and you should plan your route for each trip and take the most direct way, whilst avoiding road works and hilly terrain.

If possible, introduce flexible working hours, so that your staff do not have to travel during peak hours, where frequent starts and stops decrease fuel efficiency.

You can increase the fuel efficiency of trucks by improved diesel engines, transmissions, aerodynamics, tyres, and hybrid technologies that extend the efficiency of vehicles in stop-and-go operation and by reducing idling.

Eco-driving

How a vehicle is driven can directly impact its fuel consumption, so it pays to implement driver training or awareness programs to educate drivers about efficient driving styles that improve fuel consumption (and lower fuel costs). Eco-driving also places less strain on a vehicle, resulting in less wear and tear on the brakes, engine, and drive train.

According to EPA Victoria, fuel efficiency can vary as much as 45% between different drivers using identical cars.[128] Eco driving means:

- Maintaining an appropriate distance to other vehicles
- Conserving momentum (in gear or neutral)
- Driving smoothly
- Using higher gears
- Avoiding excess idling
- Only using the air conditioning when needed
- Using cruise control
- Avoiding congested traffic areas
- Driving more slowly – cars that drive at 90km/h save up to 16% in fuel consumption compared to travelling at 110km/h.[129]

Examples of organisations in Australia that have run eco-driving training programs include Linfox, the City of Sydney, Gosford City

Council, Telstra, Toll-IPEC, Australia Post, and the Victorian Transport Association.

Eco-driving can also be semi-automated. One example is Volvo's *I-See* predictive cruise control[130] for trucks, which employs knowledge of road topography to optimise a truck's speed and gear shifting in the most fuel-efficient way. When an uphill grade nears, the truck's speed increases, in preparation. Once the climb begins, the truck avoids downshifting, to save fuel. The truck curbs speed just before the downhill, then rolls in neutral gear downhill, avoids over-speeding by using the engine breaks, and builds up speed again in neutral gear before continuing on its journey.

Changing the fuel supply

Once you implement energy efficiency opportunities for your fleet, your medium- and long-term 100% renewable energy plans should reflect changes to the fuel supply. To make your fleet's energy needs fully renewable, you must switch to fossil fuel alternatives, like liquid and gaseous biofuels, or to electric vehicles charged from renewable electricity.

Planning for 100% renewable energy does not mean shifting everything to renewable energy overnight. Some companies choose to replace vehicles at the rate of natural turnover, when they would otherwise normally purchase new vehicles. Others develop plans that show at what point in time it makes economic sense to transition. Having a plan makes it is far more likely you will achieve your goal in the target year.

Battery electric vehicles (BEVs/EVs)

Every time I wait for my suburban train at the station and a diesel-powered, coal-filled train rattles past, I find myself annoyed by its noise and fumes as well as its payload. And once I arrive in Sydney's central business district, I am annoyed by the noisy traffic and bad air.

It is my fervent hope that the internal combustion engine is replaced by electric drive trains. I may be lucky, as it looks like electric vehicles will be the preferred technology of the future. How much better would it be if more vehicles were powered by electricity? Electric vehicles (EVs) can be charged from locally available, renewable electricity, they are quieter, cheaper to run, and maintain, and they have no tailpipe emissions.

Electric vehicles charged from the grid or other non-renewable generated electricity sources do not qualify as emission-free.

EVs are much more efficient compared to fossil fuel-powered cars, which means that the energy costs per kilometre are many times cheaper for electric vehicles. It is also much easier to change to a renewable energy source, like powering your vehicle from solar panels, once you electrify the transport vehicle.

Electric cars are not as new a concept as people tend to think. Thomas Edison built an electric vehicle in 1895, and another with Henry Ford, in 1913, which was based on the Model T frame.

However, the electric vehicle concept did not catch on until recently, when companies like Tesla showed that EVs could be sexy and high-performing, and when some countries provided more political support. Apart from Tesla EVs, there are other fully electric cars, such as the Nissan LEAF, the BMW i3, the Ford Focus Electric, and the Volkswagen e-Golf.

In 2005, the worldwide number of EVs was measured in the hundreds, whereas in 2015 EVs exceeded one million.[131] According to the IEA, annual electric car sales grew by 70% in 2015, compared to 2014.

Norway has the highest penetration (23% in 2015) of electric vehicles of any country in the world,[132] thanks to incentives like free parking, permissible driving in bus lanes, and tax benefits. In the Netherlands, the market share of EVs was 10% in 2015.

In terms of absolute numbers, China is the main market worldwide for electric cars, scooters, and buses, because of China's air quality issues. The sales of electric scooters now match petrol-powered ones after just a decade in the market, according to the IEA. In 2000, electric scooters represented a mere 1% of all scooters in China; in 2015, that number had risen to 40%.[133]

An example of a company that electrified its fleet of 10,000 light-utility vehicles is La Poste, the national mail carrier in France, which makes it the owner of one of the world's largest electric fleets.[134]

In the southern hemisphere, New Zealand Post has committed to investing $15 million in EVs for residential parcel delivery.[135] In Australia, many local governments and businesses have introduced electric vehicles to their fleet, but, unfortunately, transitioning to fully electric vehicles on a large scale is still far from mainstream.

The main barrier is the capital cost of the acquisition, which overrides the lower running and maintenance costs of the vehicles. So long as organisations do not apply the 'total cost of ownership' costing principle to their asset-buying decisions, the barrier of higher upfront capital costs remains.

Other problems are the unavailability of widespread charging points, the time it takes to recharge a car, and the recent low oil prices. Companies also struggle with the timing of the charging. Many organisations use vehicles during the day and cannot afford to have them sit idle whilst being charged from solar power.

One problem (which will soon be overcome) is the travel range of EVs. With current ranges, some people fear that their car might stop on the way with no opportunity to recharge. When I worked with a local government located in a hilly region, its fleet managers were reluctant to accept EVs as a potential option for making transport fully renewable. They argued that vehicles were unable to navigate all the hills in their area and make it safely back to their charging stations.

This will fortunately change in the foreseeable future: EVs will be cheaper, there will be fast charging stations, and a significant improvement in range. Naturally, the uptake will be even faster if there is government support.

In April 2016, Germany decided to invest 300 million euros in a nationwide network of 15,000 roadside, rapid DC battery chargers, which will push Germany to the top position in Europe in terms of rapid-charging development and enable mass uptake of EVs in the 2020s.[136] The country plans to put one million electric vehicles on German roads by 2020.

The New Zealand Government has targeted doubling the number of electric vehicles every year to reach about 64,000 by 2021,[137] which will be charged from an electricity network that gets 80% of its energy from renewable sources. In Australia, as of June 2016, Moreland City Council had developed the largest network of public EV charging stations owned and operated by a local government.

There are also heavier vehicles that run on electric drivetrains. Many cities, like Vienna (Austria), Philadelphia (USA), and Adelaide (Australia) have already introduced electric buses to their public transportation, whilst in China they already comprise more than 20% of the country's bus fleet.[138] Mercedes-Benz, for instance, has outlined plans for an electric heavy-duty truck. Its 'Urban eTruck'[139] will have a range of about 200 kilometres per battery charge and load capacity of 26 tonnes.

However, fully electric heavy vehicles are not yet readily available, so to reduce your emissions, you may need to consider switching to renewable liquid or gaseous fuels in the short term.

Biofuels

Heavy vehicles, like garbage compactors, street-sweeping vehicles, plant assets, aeroplanes, and marine vessels, need a high energy density of fuels, which is currently hard to achieve with electrification. To make these fully renewable, an option would be to power them from biofuels.

If you change your fleet to biofuels, the vehicles will still be based on the internal combustion engine, rather than an EV's more efficient electric drive train.

One of the biggest advantages of switching to biofuels is that you may not need to purchase new vehicles, as opposed to replacing your fleet with electric vehicles. All that may be required is to modify the engine to accommodate a higher percentage of biofuels.

Liquid biofuels

The production of liquid biofuels, such as bioethanol and biodiesel, reached about 95 billion litres per year in 2014, according to the International Renewable Energy Agency (IRENA). This equates to 3% of all liquid fuel used for transport.

Most of this production is due to biofuel targets and incentives that Brazil, the USA, and European Union countries have set up to diversify transport fuel supplies and improve energy security.[140]

Feedstocks for most biofuels today are agricultural crops. With these first-generation or conventional biofuels, sugar cane and maize are converted to ethanol, whilst palm, canola (rapeseed), and animal tallow are converted to diesel. First-generation technologies are proven and currently used at a commercial scale.

Second-generation biofuels use feedstocks like farm and forest residues, grasses, and trees. Some such feedstocks can have high yields, sequester carbon in the soil, and grow on land poorly suited to food crops.

Third-generation biofuels, from algae, are in the early stage of development and are not yet cost effective. However, they could grow on much less land whilst producing a variety of useful co-products.

Theoretically, biofuels are carbon neutral, as they come from the renewable versus fossil fuel cycle. However, depending on the production practice and feedstock, the creation of biofuels can be water- and carbon-intensive, which could potentially eliminate their advantages over fossil fuels.

The cultivation, fertiliser usage, processing, and transport of biofuels, as well as land-use changes like deforestation can all contribute to climate change rather than combat it. Already, biofuel crop production is a major contributor to deforestation and biodiversity loss in the developing world.

It can also compete with land for food production and increase worldwide food prices. With all these detrimental effects of some biofuels, European Union leaders negotiated a set of sustainability rules for biofuels in 2015. They also introduced a seven-percentage-point cap on the contribution of conventional biofuels towards the EU 2020 target of 10% renewable energy in transport.

According to IRENA, advanced liquid biofuels (second and third generation) may alleviate most of the sustainability concerns over conventional biofuels, as they can cut carbon emissions by 60% to 90%, compared with fossil fuels, over their life cycle, even when supply-chain and land-use change emissions are accounted for.[141]

However, advanced biofuels are more expensive to produce. As such, they currently have low production volumes, at about one billion litres in 2014, or less than 1% of total liquid biofuel production.[142]

The development of biofuels is still an emerging sector. For biofuels to be sustainable, affordable, and reliable there needs to be a considerable investment by governments, organisations, and venture capitalists, technological breakthroughs, and growth in consumer demand.

To find out more about the sustainability of biofuels you are considering, you can fill out the IDB Biofuels Sustainability Scorecard at http://www.iadb.org/biofuelsscorecard, which is based on the sustainability criteria of the Roundtable on Sustainable Biofuels.

Bioethanol

Bioethanol is a fuel manufactured through the fermentation of sugars found in grains like corn, sorghum, and barley. In the EU, bioethanol is mainly produced from grains, with wheat as the dominant feedstock. In Brazil, the preferred feedstock is sugar cane, whereas in the US it is corn (maize).

It is fairly easy to get fuel with about 10% ethanol by volume, a product called E10 (petrol with 10% ethanol). Ethanol has less energy content than conventional petrol, which means that your fuel consumption will increase by 1–3.5%[143] and at the same time be more cost-efficient thanks to the cheaper price of E10, which is the case in many countries.

Most modern petrol-powered engines can use E10, but only specific types of vehicles can use mixtures with fuel containing more than 10% ethanol. To be fully renewable, the total fuel mix has to be ethanol, with no fossil fuels.

Biodiesel

Biodiesel is a fuel produced from oil sourced from canola/rapeseed, sunflower seeds, soybeans, used cooking oils, or animal fats. If waste oils are recycled as feedstock for biodiesel production, pollution from discarded oil is reduced, providing a new way to transform a waste product into transport energy.

Biodiesel is usually sold as a blend of biodiesel and petroleum-based diesel fuel. Most countries use a labelling system to illustrate the proportion of biodiesel in the fuel mix. Fuel containing 20% biodiesel is labelled B20, whilst pure biodiesel is referred to as B100.

Blends of 20% biodiesel with 80% petroleum diesel can be used in diesel engines without changing the engine. B100 can be used in some engines built since 1994 with biodiesel-compatible material for certain parts, such as hoses and gaskets.

B100 has a solvent effect, potentially cleaning a vehicle's fuel system and releasing deposits accumulated from petroleum diesel use. Note that the release of these deposits may initially clog filters, and necessitate frequent filter replacement in the first few tanks of high-level blends.[144]

To be classified as fully renewable, you must use B100 or a pure, ethanol-based fuel.

Gaseous biofuels

Biogas

Gas is widely used as a transport fuel in many European countries, notably Italy. For gas-powered vehicles to be fully renewable, the gas must come from biomass. Biogas is derived from decomposing organic matter in a digester. (See Chapter 4 for a more detailed explanation.)

Biogas cannot be used in its raw form – first, it must be cleaned to remove impurities and the methane enriched and compressed. This means that biogas production is capital intensive and most likely not located at the point of demand, like a depot. Whilst vehicles exist that can be powered from biogas, there is not enough biogas available and a lack of an adequate refuelling infrastructure.

Despite all these barriers, Sweden has emerged as a world leader in upgrading biogas and using it for transport. The country has many biogas vehicles, including private cars, buses, and even a biogas train.

Hydrogen (fuel cells)

In the future, it may not be possible to power all vehicles from rechargeable batteries or solely by biofuels. Hydrogen has the advantages of a high energy density, highly efficient fuel cells, and a lighter payload. Also, refuelling takes only a few minutes, which enables heavy or larger vehicles to travel longer distances.

Like battery electric vehicles (BEVs), hydrogen-based ones have an electric drive train and do not emit any greenhouse gases during operation. When burned, all that is released is water and heat. Their lifecycle greenhouse gas emissions, like with liquid biofuels, depend on how hydrogen is produced. For hydrogen to be a truly renewable resource, several conditions need to be met:

- It has to be produced via electrolysis, rather than extracted from natural gas
- The electrolysis must be powered by renewables like wind, solar, or hydro plants
- The energy required to convert the gaseous hydrogen to a liquid hydrogen for easy storage needs to be produced from renewable energy sources
- The transport of hydrogen from the production plants to the service stations must happen with a renewable transport fuel – the same, of course, holds true for the transportation of biofuels.

Other barriers to overcome revolve around fuel-cell cost and performance, on-board hydrogen storage technology, and the refuelling infrastructure. Examples of car manufacturers that developed hydrogen fuel-cell cars include Mercedes-Benz[145] and BMW.[146]

Purchasing carbon offsets

The goal to achieve 100% renewable energy for your transport needs can be challenging, especially in the short to medium term. You may not be able to fully electrify your fleet at a given time, or have all of your fleet run on renewable fuels. You may also not have renewable fuel supplies readily available, or at the quality you need.

If you include transport in your boundary definition of 100% renewable energy, and depending on how far out in the future you must meet your goal, you may need to purchase carbon offsets in the short term. You can find out more information about carbon offsets in the last section of the previous chapter.

Your checklist:

The following list only applies if transport energy is part of your boundary. You may be able to delegate or outsource these tasks.

Examine, on a high level:

- ☐ Your light and heavy fleet, and how you could reduce fuel consumption
- ☐ Your light and heavy fleet and the readiness and availability of renewable energy options
- ☐ Whether you need to buy carbon offsets to achieve 100% renewable energy for your transport energy needs.

Chapter 6

Battery storage and other emerging trends

The innovation that has arisen in the past years has been nothing short of amazing. You only have to turn the clock back to the end of the previous century to see this: no smartphones or tablets with apps connected to the internet, no YouTube, no Facebook, no Netflix, no affordable 3D printers, no computers that could fit into your pocket, and no location-based services like Uber.

The mobile phone you likely carry in your pocket boasts more computer power than all of NASA in 1969, when it placed two astronauts on the moon. The Sony PlayStation, which costs about $300, has the power of a 1997 military supercomputer, which cost millions of dollars.[147] Technology advancement is not linear but exponential.

In 2000, electricity was cheap in Australia, and many organisations were only beginning to notice how much energy they used. Nowadays, there is a much bigger focus on how much energy is being consumed, and many appliances and equipment have become more energy efficient.

Many organisations have set themselves carbon reduction targets and work hard to achieve them. Tesla has captured imaginations, and people are making down payments on its Model 3 car that, as of 2016, isn't even available yet.

We used just to consume electricity; now we produce and even trade it.[148] We once shovelled coal into our power plants; now we increasingly harvest energy from the sun, wind, water, and biomass.

With all the changes since the turn of the century, it's easy to wonder what kinds of changes will occur in the next ten or fifteen years, which is about the time frame organisations set themselves to achieve a 100% renewable energy goal.

We tend to seek solutions based on existing technology ideas, which risks missing a longer-term vision most likely based on a much different reality. Put the clock forward to 2030. What will our power generation and consumption look like then? How renewable will the grid be? How will the Internet of Things change the way we operate our energy assets? What are the implications and opportunities for your 100% renewable energy target? What will future changes mean for the achievability and the timing of your goal?

Disruption and change happen so rapidly that it is hard to make predictions. However, most likely, there will be more (not fewer) opportunities available to you, which may include battery storage, virtual net metering, peer-to-peer energy trading, 'smart' grids, and vehicle-to-grid technology.

Battery storage

Unlike other forms of energy, electricity, once generated, has to be consumed immediately. Some people, when they think about electricity from the grid, assume that the generated power is stored somewhere for us to draw upon when we need it. However, this is not how the grid works, as the supply of electricity needs to be balanced with the demand at all times.

Introducing storage allows us to disconnect consumption from generation and use the energy at a later time. Some mountainous regions have been lucky to be able to use the traditional method of storing energy by using pumped storage hydro plants,[149] utilised for more than a century.

However, the focus of this chapter is on battery storage, which – once it becomes cost effective – will help with the widespread adoption of distributed, renewable energy resources. Batteries have been around for a long time, and even Thomas Edison at the beginning of the twentieth century used them in cars and for stationary purposes. His vision was to connect batteries to windmills, making remote households entirely independent from the grid.

I am a big fan of the freedom that batteries provide. I still remember my excitement when I first used my Walkman. And, of course, now, as I write this book, my laptop battery allows me to write nearly everywhere. In terms of renewables, though, batteries enable us to turn wind and solar electricity into a 24/7 power source as reliable as baseload power, based on coal.

While batteries have existed in torches, kids' games, cars, and many more things for longer than a century, they have only just begun to take off as a technology for energy storage. Electric cars are growing in popularity, and the automotive battery supply chain is driving the transformation in stationary storage. Battery safety has improved, and costs have decreased significantly in the past years.

With all this progress, battery storage is gaining more and more traction in the market, with the future looking even better. A 2014 report by investment bank UBS predicts a 50-fold increase in global energy storage technology by 2020.[150] And if governments decide to subsidise the installation of battery systems, the uptake will grow even further.

It will also get easier to finance battery systems. Just as there exist innovative financing solutions for renewable energy installations, there are similar options for battery storage. You might be able to get a battery system installed at your site for zero upfront dollars. The installer company would retain ownership of the battery system and take a portion of the savings it generates.

Australia is seen as a test bed for battery storage technology innovation, with its relatively high electricity prices and the largest penetration worldwide of rooftop solar PV.[151] Not surprisingly, a lot of international battery storage providers are setting up offices in Australia. Disadvantages of batteries include their limited lifetime, as you can only recharge them for a certain number of cycles. They can potentially be dangerous if not treated appropriately, which can lead to fire, explosion, or chemical pollution, which is why they need to be maintained and checked regularly and should be installed in a secure, fire-rated room. They are also sensitive to their environment. For instance, if they are installed in extremely hot or cold climates, their long-term performance may drop, or they may stop working altogether.

Battery storage systems may contain heavy or toxic metals harmful to the environment if disposed of in landfills, making it important to consider the end-of-life recycling of the batteries.

AC- versus DC-coupled battery storage

Power from the grid is *alternating current* (AC) electricity, which is the power available from wall sockets. However, the power that solar panels produce and batteries hold is *direct current* (DC) electricity. This is why you need an inverter to convert from DC to AC electricity, both for the power your solar panels generate and that which your battery system stores.

If your battery system is DC-coupled, it can share the inverter with the solar panels. If it is AC-coupled, it will require its own inverter. An AC-coupled battery system can store excess energy not only from your solar panels but also from the grid, which is ideal for tariff arbitrage purposes (to be explained in greater detail further on). An AC system usually also includes an *uninterruptible power supply* (UPS), which can provide you with backup power, should the grid fail.

DC-coupled battery systems are typically cheaper to install, but unable to store grid power – they can only store the DC power generated by your solar panels. DC systems are also unable to provide backup power, as the system must shut down for safety reasons when there is a blackout.

Management software

Installing batteries is about more than just purchasing a battery. If you get a system with an inverter, you also get an energy management system, often separate to the inverter, which shows and controls what is being generated, consumed, and stored.

The system needs to be tuned so that everything can run in the most cost-effective way, which for most organisations is to:

1. Buy electricity when electricity prices are low (off-peak)
2. Self-consume when prices are moderate or high (shoulder, or peak)
3. Sell excess electricity when electricity prices are high (peak).

The management software tells the system when the power should be stored versus used versus exported (sold) to the grid. For example, on a sunny day, when your solar PV panels generate more electricity than you need, the power goes to the battery, to recharge it. Once the battery is fully charged, the excess power might be sold to the grid. Then later, when the sun starts to set and you need more energy than your renewable energy system is producing, the battery discharges electricity to power your operations.

Clever management software also integrates with building management software. This allows you to monitor and manage real-time and historical performance metrics, like building load, battery charge and discharge rates, your renewable energy production, and the state of the battery charge.

The applications for batteries

Households will benefit first from battery installations, as they face higher electricity prices than commercial users and their peak consumption happens mornings and evenings. However, with falling prices, battery storage will become important for organisations as well.

For most business scenarios, the current reality is that batteries may take longer to pay off than manufacturers give a warranty for. However, this will change in the next few years. The business case for batteries (especially in conjunction with renewables deployment) will be better in the next few years, due to these driving factors:

1. Costs to generate power from solar PV and wind have, or are reaching, grid parity and users want to self-consume more of the power output
2. Prices for batteries continue to fall
3. Financial incentives to feed excess electricity into the grid are becoming more limited
4. New electricity network pricing structures may be more expensive during peak times[152]
5. Solar PV providers have started to package batteries into their offer
6. Companies like Tesla are driving rapid innovation.

The business case for batteries is already viable in some cases – for example, where energy supply is unreliable or expensive. Island communities, for instance, that once imported diesel at high prices to run power generators, have already started to use batteries to switch to renewables instead.

As shown in the next graphic, apart from remote locations, and depending on the unique circumstances of each business, the next applications that may become viable, in no particular order, are:

- Demand and time-of-use charge reduction/peak demand shaving
- Increased self-consumption of renewables
- Backup power.

It is important to note that each of the following applications is different and complex, and it is advisable to seek proper professional advice to help you select the best solution.

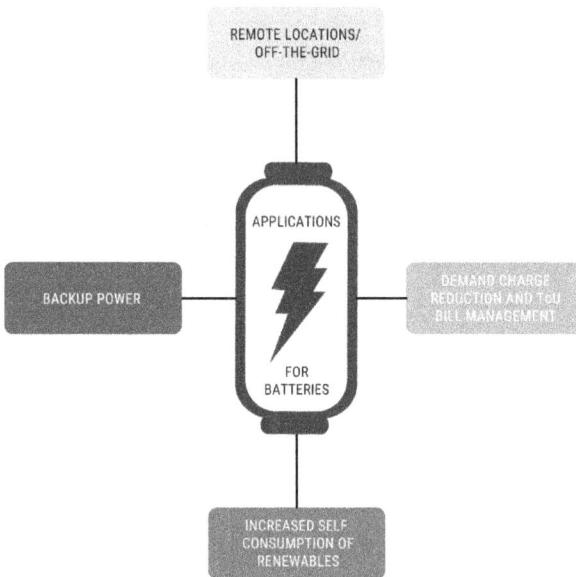

Figure 18: Applications for battery storage

Remote locations and off-the-grid applications

Remote areas represent one of the most attractive opportunities for battery storage, as supplying electricity (or diesel) to remote locations is expensive. With renewable energy solutions and battery storage constantly getting cheaper, renewable energy is already the preferred source of power for many mining and remote towns in Australia.

Tokelau, an island in the South Pacific, north of Samoa, was once powered by diesel, which was expensive and noisy. In October 2012, Tokelau became the first place on Earth to be entirely powered by solar photovoltaics with battery storage.

In Australia, most demand for off-grid renewable energy comes from remote locations, like pastoral stations, rural and indigenous communities, tourist facilities, small industrial projects, pumping and irrigation, and mine sites.[153] The town of Tyalgum, for instance, located on the fringe of the grid, is working on plans to voluntarily disconnect from the grid with the help of battery storage and renewables.

For other businesses located close to the grid infrastructure, it will be cheaper and less complex to stay on the grid. Going off the grid is expensive relative to remaining on it. You also need to size your battery system sufficiently to provide power so your business can still run, even if your renewable energy systems do not produce power for some period. This is especially the case if your organisation is powered by intermittent renewable energy sources, like solar, or wind.

> Decisions to stay on the grid or disconnect and use renewables are not simple and should be based on an appropriate cost-benefit analysis.

However, in certain circumstances, it might be cost beneficial to have a portion of your operations go off the grid. An example of an organisation achieving this is the North Coast Institute of TAFE NSW, a tertiary training institute covering a broad area along the east coast of Australia.

The institute's primary market segment is the training and upskilling of tradespeople, but it also partners with organisations that need to provide training for their workers. As part of a recent highway upgrade along the coast, many people needed to be trained in civil construction.

It was not feasible to get all the students to a central location, so the institute innovated and created self-sufficient classrooms, called PODS (portable onsite delivery spaces), customised containers that can be shipped on a truck to wherever the training needs to take place. The pod is completely stand-alone in water needs, as well as in energy through the help of solar PV cells and battery storage systems.

The institute partnered with a local energy storage system provider which saw this as a great opportunity to test the application of its systems and gain popularity for its products. The financial payback period for the battery component of this project was just over six years.

Reducing peak demand charges and time-of-use bill management

Batteries already make sense for some off-the-grid situations, but in future years they may become profitable for tariff arbitrage, like peak demand and time-of-use management. Generally speaking, the higher your electricity tariff, the sooner the business case for battery storage will be viable.

If you are a large energy user, peak demand charges may make up a significant part of your organisation's electricity bill, sometimes even half. These charges are based on your peak usage during 15- or 30-minute intervals in your billing period.

According to the Australian Energy Market Commission, new electricity network pricing structures will reflect the true cost of energy.[154] Such pricing structures will provide a greater incentive for your organisation to reduce your energy use during peak electricity use times when costs are highest.

If your renewable energy system is big enough to generate more electricity than you need, you can feed the surplus power into your battery system, rather than exporting it to the grid. If you do not produce excess electricity from your renewable energy system, you can use an AC-coupled battery to benefit from the different tariffs that may apply during different times and days of the week.

You could, for instance, charge your battery from the electricity grid during the off-peak time to take advantage of low-cost, time of use, and economy tariffs. You could then use the stored power during peak times or to reduce peak demand charges. Depending on how your battery system is set up, you might be able to prioritise your battery storage system so that it charges first from solar and then from the grid, or strictly from solar.

Whilst economically beneficial, charging your battery from the grid during off-peak times will not necessarily reduce your emissions, as grid-supplied electricity is not fully renewable.

If your battery is fed by off-peak grid energy (rather than your renewable energy system), you can size the battery based on your daytime network demand, electrical capacity, available space, and required financial return.

Increased self-consumption

With batteries, renewable power can become available 24/7. Instead of matching the solar PV size to your site's load, you would oversize the system. Oversizing *without* battery storage means exporting a lot of electricity to the grid with low financial return.

However, *with* battery storage, the system captures surplus energy generation and deploys it when solar generation is small or non-existent. This means the storage solution fills the gap when the sun does not shine. This business case is especially applicable to organisations with a high demand for electricity at night, such as clubs or entertainment venues, and also businesses with a high early morning demand, like bakeries.

As with the other business cases, the more you pay for your electricity, the sooner the business case for battery storage becomes viable.

Backup power

Backup power is a top-of-the-mind item for many organisations that have been affected by severe weather events, like floods or superstorms, or for organisations that provide critical infrastructure. It is also important for those whose continued operation is critical in the event of a power failure.

I was once consulting to a food producer with two power supplies and big backup generators, in the event they lost both power supplies to the site. Losing power would be disastrous for their equipment as it could potentially be destroyed if the production process stopped, with heated ingredients sticking to the walls of the process equipment and then slowly cooling, resulting in the destruction of their assets.

AC-coupled batteries can increase energy security in case of grid failures. Rather than relying on a backup generator powered by diesel, organisations can consider installing battery banks instead. Batteries, combined with renewable energy generation (and potentially smaller diesel generators),[155] might provide adequate energy security.

If you choose batteries for backup power, you will pay more for your system compared to solutions where the batteries shut down when the grid fails. This is because the inverters must disconnect from the grid and control electricity supply and demand. There will also be more wiring work required.

If you want your battery to act as a backup during times of power outages, you need to analyse your load carefully and consider the minimum load to be supplied from the battery, so that you can continue running all or just essential parts of your business.

To size the battery system correctly, you need to calculate how long you want your system to provide you with backup power. Should it be for two hours, six hours, or even 48 hours? Maybe you only need the backup power to be able to do an orderly shutdown of your production processes, or maybe you need your full business to run because you need to be in a position to provide essential services to your customers.

To be able to run your battery storage system as a backup service, the batteries would also need to be at or near full charge.

Other emerging trends

Vehicle-to-grid

Plug-in electric vehicles are a potentially viable alternative to stationary electrical storage. Cars are parked 95% of the time,[156] so they could possibly be used to balance demand spikes on the grid. This will enable your organisation, as the electric vehicle owner, to sell excess energy stored in your vehicle batteries to the grid during times when the network is experiencing peak demand.

Selling the energy will generate income for your organisation whilst it helps balance demand and supply on the grid. Technically, this solution is possible – it is more a question of matching the charging and discharging to driving habits and making sure you do not violate your battery's warranty.

Virtual (remote) net metering

Net metering is when you generate renewable energy behind-the-meter and receive a credit for the renewable energy you export to the grid. Any renewable electricity you do not consume but instead export is called 'net excess'.

Virtual/remote net metering extends the concept of net metering to outside your site. Under this type of net metering, you can assign your onsite-generated energy to another site.

The term 'virtual' is used, as the exported electricity is not physically transferred to the other consumer, but for reconciliation purposes, is billed to them. In the UK, several states in the US, and in Germany, small renewable energy generators legally sell their electricity directly to one or more nearby electricity users.[157]

Currently in Australia, if energy is exported to the grid, the full network costs are incurred, no matter over what distance you export the electricity. So, even if you want to export electricity to a site a few hundred metres away, you would incur a big network cost to use the poles and wires in between.

One option to solve this problem would be to lay a private wire between the two sites, but this is expensive and could duplicate existing infrastructure.

One of the local government clients I consulted to has a large administration building that consumes a lot of energy. Its roof is shaded by palm trees and the aspect is not ideal – only a small solar PV system could be fitted onto it.

On the other hand, their aquatic centre was just across the road. The swimming pool had an enormous roof space which could hold a large solar PV installation. The maximum size of PV possible, with the given dimensions of the roof, would exceed the demand of the aquatic centre.

With virtual net metering in place, the council could put the maximum possible size of solar PV on the roof of the aquatic centre and then export the excess electricity to the administration centre across the road; however, with the current network pricing structures, the business case for doing so was not viable, as the incurred network charges would be cost prohibitive.

The argument from the end user's point of view was that the electricity only has to travel a short distance before it arrives at the next place of use, making it seem unfair to pay full network costs to transport power between two adjacent sites. If, instead, only a small portion could be paid to the network provider, the business case would make sense.

At present, virtual net metering is not widespread in Australia, but there are a small number of trial projects, like one in Byron Bay.[158] If they are successful, new tariffs may be introduced that provide a discount to the standard 'Network Use of System' tariff.

What this means for your organisation is that your solar PV systems could potentially be oversized to exceed the local demand of your site, the excess energy exported to the grid and credited to another site close by.

Peer-to-peer energy trading

Most of the world's energy today comes from large, centralised power plants. The renewable energy revolution will not follow this centralised, top-down approach.

Instead, it will probably look like the internet or telecommunications revolution by embracing distributed business models. Today, if your organisation wants to sell excess power from a renewable energy installation to the grid, you have essentially two options: you either sell to your retailer or you do not sell at all. You also have to accept the price the local utility offers you for the exported electricity, which is usually the wholesale price for electricity, or less.

Peer-to-peer energy trading is similar to virtual net metering. However, rather than involving the retailer to credit the generation of electricity from one location to another, you sell your excess renewable electricity directly to your neighbours.

A typical organisation operates five days a week, which means that on weekends renewable energy systems like solar or wind produce excess electricity that is not needed. Conversely, more people are at home during the day on weekends.

Theoretically, the excess energy could be sold or attributed to users down the street. An example of what this could mean for an organisation is their ability to 'donate' electricity to disadvantaged communities during times when they, themselves, do not need the power.

Currently, there is no mechanism to assign renewable energy to neighbouring properties, other than to export it to the grid at low prices. This is despite the fact that – *physically* – the electrons already flow there.

Peer-to-peer energy trading will be similar to current P2P file-sharing services, like BitTorrent. Organisations with a 'smart' meter can join the market as a buyer or seller of renewable energy. They set their buy and sell rates for renewable energy via an online portal, with the smart meter tracking how much energy is imported or exported.

To enable peer-to-peer energy trading, the energy market needs to be deregulated and the necessary IT and accounting infrastructure in place. Currently, peer-to-peer trading is not available in Australia.

Peer-to-peer energy trading might mean electricity will be cheaper than that offered by local retailers, which in turn might weaken the business case for renewable energy generation and storage. Conversely, it might help large-scale energy producers by offering electricity at lower rates, as the retailer margin is removed from the equation.

Smart grids

A smart grid means that the electricity grid will be computerised. Rather than energy flowing from power generators to consumers, the grid has a two-way digital communication technology added to it that allows devices to communicate with network operators.

Smart grids create value up and down the value chain,[159] much like the internet transformed our lives with swift, universal communication that enables sophisticated transactions and creates new business models. Smart grids may allow similar things in the energy market with a two-way network that is flexible and secure.

Smart grids are self-healing and can rapidly detect, analyse, and respond to problems and restore service quickly. They are more resilient to physical and cyber attacks, and can accommodate a wide variety of local and regional renewable generation technologies.

For smart grids to work, energy end users need smart meters installed. With smart meters you can see near real-time data on electricity prices and your consumption, allowing you to make better decisions regarding energy usage. They also enable you to participate in demand reduction actions during periods where the grid experiences peak demand and receive a monetary reward for your energy reduction.

With smart meters, automation technologies, and your consent, network operators can tell energy-intensive equipment, like air conditioning, to throttle down, or not to run for a brief period. From your organisation's perspective, it can mean cheaper electricity rates through dynamic, real-time pricing, and you will be rewarded for reducing the demand on the grid.

From a utility's perspective, smart grids also make the integration of intermittent renewable energy sources and electric vehicles into the grid easier.

Microgrids[160] are a subset of smart grids and will play a major role for communities that want to switch to 100% renewable energy. They are localised, distributed electric grids that can disconnect from the larger electric grid, and are ideal to enable the local generation and consumption of renewable energy. They can be used to power a single facility or a larger area, and can extend the boundary of behind-the-meter installations.

Nanogrids are a subset of microgrids and typically serve a single building or load. Nanogrids are conceptually similar to electricity systems in vehicles or Power over Ethernet (PoE) distribution systems, which pass electric power, along with data, on ethernet cabling. They are often built for direct current (DC) solutions, which can be connected to the grid or be stand-alone. The advantage of using a DC solution is that conversion losses, as well as investments in inverters and breakers, decrease. Nanogrids will also have device-level controls that enable a better way to match generation or storage capabilities to the demand.

Internet of Things

The Internet of Things (IoT) is a network of physical objects that produce and communicate data with each other over the internet, creating the opportunity to evaluate and act on the information. The energy sector is already applying this concept with smart meters, demand response systems, smart thermostats, and smart lighting.

However, in the future, sensors will be placed in many more physical objects, such as production systems, heating, ventilation, and air conditioning (HVAC) equipment, and solar panels, to collect data. Robotics and advanced analytics will be capable of quickly analysing large amounts of data and applying changes in the system.

What this means for your energy systems is that they can constantly optimise themselves, without the need for human intervention. In buildings, for instance, HVAC systems will be more efficient, lights will automatically be turned off when a room is not occupied, airflow will be balanced so that occupants get air where and when they need it, and the temperature will be more efficiently controlled, responding to the demands and requirements of the people inside them and the weather outside.

The IoT will be vital for creating smart grids, in which a huge amount of data from buildings, vehicles, power generation assets, and energy consumers will be interconnected via smart meters. The IoT will also help make our cities 'smart'. Traffic lights and jams will be rare, even non-existent, and you will save on fuel and energy costs because everything works so efficiently.

What the future holds

The energy industry is undergoing a digital, regulatory, and business model transformation, and once this has happened, the energy space will look vastly different from the one we have today. In as few as five years from now many things will change, some of which we are not even close to foreseeing.

Solar that produces electricity and/or hot water will be integrated into roof or façade materials, installing renewables will be the standard way of doing things, and energy start-ups will provide us the necessary additional add-on services to revolutionise this space further.

The grid will accommodate distributed energy, be digitally enabled, and become a two-way street, with electrons flowing from generators to consumers and from 'prosumers' to other consumers. Organisations may trade excess renewable energy produced onsite with their neighbours. Utilities will introduce new price and tariff structures to accommodate the distributed generation and storage that organisations implement.

In the future, the majority of vehicles will be powered by electricity, and others by biofuels, which will not only allow us to produce energy locally but also give us cleaner air. Liquid biofuels may be based on algae, and some of our vehicles might even be powered by hydrogen. There will be fewer vehicles owned and more shared.

Thinking about these developments makes me feel excited to be involved in making our energy use more sustainable and to see first-hand how the developments unfold, for a better, cleaner way of doing business.

Now that you have all the pieces of the puzzle, you can start developing a strategy that builds the pathway for your organisation to transition to 100% renewable energy. The next chapter will lay the groundwork for Step 2, 'Plan', which is a thorough analysis of your organisation's energy situation.

Your checklist:

You may be able to delegate or outsource these tasks.

- ☐ Examine whether battery storage will be part of your solution to achieve 100% renewable energy on a high level.
- ☐ Investigate whether in your location you have access to virtual net metering or peer-to-peer energy trading and whether this could be part of your pathway.
- ☐ If the innovative options above are not yet available, liaise with your energy retailer and network provider.
- ☐ Keep emerging trends at the back of your mind when you develop your pathway. They could form part of your long-term action plan.

Step 2 – Develop the plan

At this stage in your journey, you have committed to a target of 100% renewable energy and are aware of the current and potential future opportunities available. Now you need to start planning the transition by matching these opportunities to your situation.

Unless you develop an overarching strategy and supporting implementation plan, you will not know where to start, or what to do and when. There is also a danger that you will be unable to clearly communicate your approach to the rest of your organisation. As a result, you may implement ad hoc projects but your efforts may lack coherence, and may not yield the best results possible.

To meet your goal by the target year, it is essential that individuals and teams take responsibility for the actions to be implemented, and it is important that a main person drives it. I strongly advocate spending considerable time on plan development, focusing on costs, time frames, and deliverables, so that these roles and responsibilities can be developed alongside the actions your organisation undertakes.

Your plan will provide you a roadmap and the signposts for your journey to 100%. Done well, it will show you the most cost-effective way to meet your goal and – if communicated properly – have the buy-in of critical stakeholders. Having a plan also enables you to implement concrete measures that get you steadily closer to your goal rather than wandering through many trial and error projects, or stop/start initiatives.

I recommend you start your planning process by investigating your current energy situation and what it will likely be in the target year. For this, you need to project your energy situation into the future, which I will talk about in the section 'Projecting future demand'. Once you have a complete picture, you can plot how much closer your energy efficiency and renewable energy projects will get you to your target, how much they will cost, and how much you will benefit from their implementation.

You can choose opportunities by looking at their technical requirements, their feasibility and cost, and how you can deliver and finance them. You can also look at your existing in-house skill sets, and how each potential opportunity ties in with the broader objectives of the organisation.

You must also identify your most relevant stakeholders and bring them along on the journey so that the opportunities align with your organisational objectives, overall goals, and your culture. It is particularly important that you receive endorsement from senior levels, like executives, the CEO, and/or board members.

Having stakeholders participate in the planning will shape and refine the solution and address any concerns or barriers early on in the process. It will also result in less resistance to the changes through a greater sense of ownership, and provide you greater insights into the feasibility from a broader range of people.

The end result of the planning phase should be a fully developed pathway and action plan, which contains the most suitable energy opportunities. The strategy will have the buy-in of your stakeholders, be clearly communicable, and your organisation will be ready to implement the plan in a smooth transition to 100% renewable energy.

Chapter 7

Understanding your energy profile

Having a proper grasp of your energy profile helps to understand the scope of the challenge that lies ahead. You need to measure your baseline and project it into the future before you can develop your pathway. And you need to analyse your energy balance and load profiles in depth, and organise energy audits, unless you have undertaken them previously.

The baseline

The baseline is your current energy situation regarding consumption and expenditure, energy sources, existing electricity and fuel contracts, and the tariff structures that apply to your energy supplies. Establishing this baseline enables you to compare your energy performance before and after you make changes.

To understand the baseline, you need to analyse the energy information of your facilities and, if transport is included in the scope, your fleet. This will be an easy undertaking if you already track your energy use and have experience analysing this sort of information.

If you do not have this data available, I would encourage you to establish your current consumption profile. You can either outsource this task to a consultant that will do most of the work for you, or you can gather the data yourself. You can go through your utility bills, contact your energy retailers, query (or install new) meters, or extract reports from your building management system (BMS). This process, in itself, can identify existing inefficiencies and duplications in your energy expenditure.

Your data collection system does not have to be sophisticated. It does not matter whether you track and monitor your energy consumption with simple spreadsheets or with commercial-grade software, so long as you have the energy information available.

When you establish your baseline, pay attention
to the following questions:

- How reliable is your current data collection?
- Can you trust that the information is correct?
- Are you missing a few energy accounts that might be paid directly by other departments?
- Has the information been correctly extracted from the underlying data or bills?
- Does the retailer provide you with accurate information or are you paying for an account you thought had been disconnected years ago?
- Does the fleet department supply you with correct information on fuel consumption?
- Is monitoring in place so you can see how much is currently generated by your existing renewable energy systems?

The further back your energy data goes, the better you can plot your energy consumption trends over time. One year would be the minimum amount of data you need, and three years would be very good. Analysing past performance gives invaluable insights and shows whether your energy use has increased or declined over the years. For instance, your organisation might have grown, which would result in higher power consumption. Alternatively, you might have divested some business units, which may have reduced consumption.

You can also analyse the individual trends for your various facilities, as one or more might have different growth or decline patterns and be at different stages in becoming more energy efficient or having renewables developed onsite.

In analysing your past performance, you may see the effect of energy projects previously undertaken. Perhaps you upgraded your lighting, or retrofitted fans and motors with variable speed drives.

With all other things being equal, energy projects should have reduced your consumption and your associated carbon footprint. If your company has grown over the years, but you have implemented efficiency measures, or your operational processes have improved, your overall consumption might have stayed the same, or it may even have declined. Possibly, you have current energy projects under way, the effects of which you will only see a bit further down the track.

The following bar chart shows an example of one of my local government clients that, after growth in consumption from Year 1 to Year 3, managed to reverse that trend and decrease demand for three consecutive years. However, due to growth in the council's services provision, it experienced an understandable increase in consumption again, in Year 6.

As this particular client included transport energy in its 100% renewable energy target, its baseline not only shows the electricity consumption of facilities, street lights, and gas but also the fuel consumption of its light and heavy fleets.

To display all these energy sources in one graph, convert your electricity, fuel, and gas consumption into one standard unit, like kWh or megajoules.

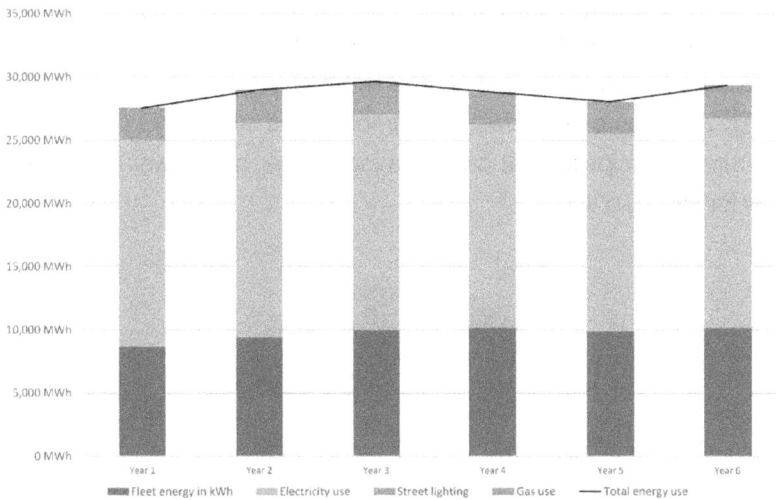

Figure 19: Example of an energy baseline for a local government organisation

> To make your current energy consumption more tangible, look up the median consumption of the average-sized household in your area[161] and calculate how many households' worth of electricity you consume annually.

If you do not have data going back in time, take the most recent year as your baseline. If you have carbon reduction goals, in addition to your renewable energy target, you should establish your carbon footprint baseline. This will be different from your energy baseline, as the carbon footprint contains more emission sources than just your energy consumption.

Examples of this would be emissions from refrigerant usage, business travel (like flights and taxis), freighting goods, use of items like paper, or methane emissions from landfill sites resulting from waste. The National Carbon Offset Standard guide[162] provides useful information to develop a carbon inventory, as does the GHG (greenhouse gas) Corporate Standard[163] and ISO 14064–1:2006.[164]

To further analyse your baseline you should also identify where most of your organisation's energy is consumed. Pie and bar charts are great visualisation tools to show management where the biggest guzzlers of energy are. Figure 20 shows an example of the electricity consumption of a rural municipality that owns the local water and sewer assets and pays for street lighting.

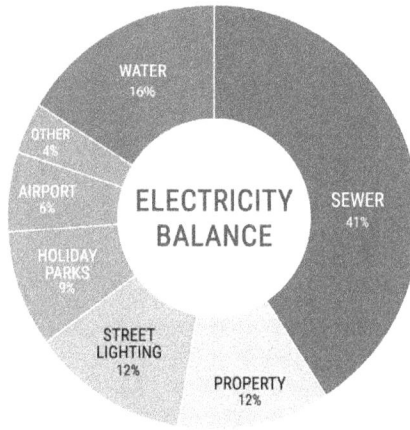

Figure 20: Electricity balance of a local government organisation

It is typical for a regional council's highest consumption to be in the energy-intensive water and sewer divisions, which is usually followed by property and street lighting. This particular local government also owns holiday parks and the local airport.

The next question you need to ask yourself is what input sources currently contribute to your energy consumption. If electricity, transport, and stationary fuels are included in your target, your most likely supply sources could be any or all of the following:

- Electricity supplied via the grid (which is usually a mix of fossil-based and renewable energy)
- Electricity and/or heat supplied by existing renewable energy systems, like solar PV, or solar hot water installations

- Renewable energy purchases
- Natural gas
- Liquefied petroleum gas (LPG)
- Biogas
- Diesel
- Petrol
- Ethanol blends
- Biodiesel blends

Your current baseline might be entirely made up from fossil fuels, or there might be a percentage of renewable energy supplied from sources on- or off-site. One of the most common renewable energy installations you may already have is solar PV panels on your roofs.

You can plot the various supply sources, like electricity from the grid, or power supplied from solar panels, onto the same style graph used in Figure 19. Not only does this enable you to see the current distribution of the energy *demand*, but also the current supply. You can see what such a graph looks like in Figure 21.

To give you a complete baseline picture, not only is it necessary to collect hard data but you should also talk to people at different organisational levels to gain more information about current management practices and behaviours. You may find that certain sections of your organisation have a good visibility and understanding of their current usage, and other sections considerably less so.

You need to identify the strategic drivers for the organisation and any future plans. If available, you should research section business plans and find out whether there are any existing feasibility studies, asset reviews, audit reports, or other previously done energy/carbon work.

Once you identify your total energy consumption for the past year(s) and study the existing relevant reports, you will have a much better idea of your starting point. This will form the basis for any predictions of what your future energy consumption may look like.

Projecting future demand

Why is it important to project demand? Firstly, you have probably set your target for some time in the future, so you need to meet your energy consumption with an equal amount of renewable energy production at that point in time.

Secondly, your future energy consumption will most likely differ from your current one. Your organisation might grow, or it may reduce its operations. Changes in the size or production output of your business typically mean a change in energy consumption. However, these changes might be offset with energy projects or other asset or technology upgrades already underway or planned.

And, thirdly, projecting future energy demand will help to quantify your targets, which means you will know how much energy must be supplied from renewable sources in the target year so you can be confident of meeting your goal.

The best way to make a good prediction is by talking to a lot of people across different divisions in your organisation. Set up meetings with senior management to find out about the various market forces that influence your company.

Talk to site managers and learn what future developments are planned at the site.

- Do they think the site's demand will grow or decline? By how much?
- What energy measures were implemented in the past?
- What actions are planned?
- What energy efficiency outcomes do they plan to incorporate into new projects?
- Do they envisage the energy profile of the site changing significantly due to, for instance, a complete overhaul of the existing operational processes or equipment?

The result of your work should be a table or a graph similar to the one in Figure 21 below. The way this chart was completed was by applying a steady annual growth of 1.5% in energy consumption. The 1.5% was not chosen arbitrarily but instead estimated based on the projected growth rate of the population in the municipality, as well as information on future asset upgrades affecting energy use.

More people in the local government area means more services must be provided, which most likely results in an increase in energy consumption. The business-as-usual scenario is a prediction of what will happen without any further efficiency and renewable energy measures.

The dark line in the picture represents a path to 100% renewable energy and shows the kinds of reductions that need to take place to move closer to the target. In this particular case, the organisation chose an interim goal of a 50% fossil-fuel reduction in 2025. Your actual journey might not follow this smooth curve. Instead, your energy projects might decrease your fossil-fuel consumption in step changes.

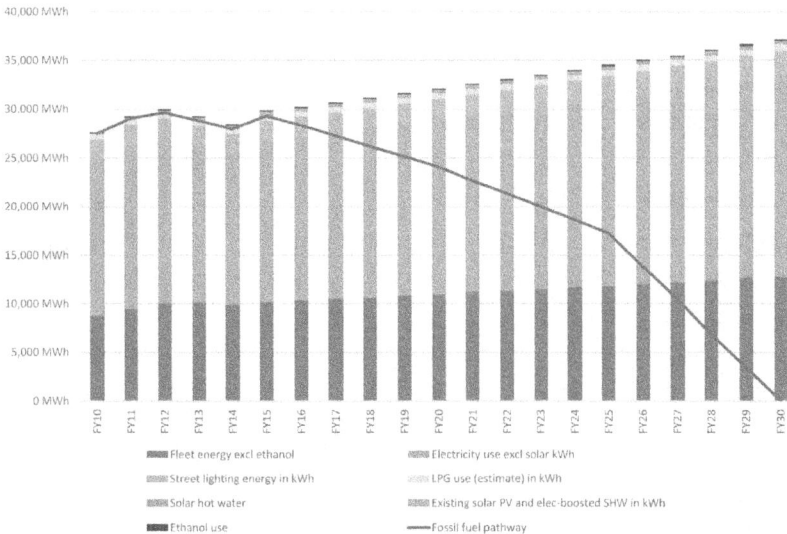

Figure 21: Example of an energy use projection for a local government organisation

If you plan to switch your fleet to electric cars, do not forget to factor in growth in your electricity consumption and a decrease in your transport fossil fuel usage when you develop your pathway.

Having the current and future energy consumption and what renewables contribute to the overall total will give you a better understanding of the journey that lies ahead, but you need to find out more about what specific energy efficiency and renewable energy opportunities you have. Some of the tools that can help you with this are load profiles, energy audits, and discussions with key operational staff.

Load profile

A load profile illustrates your daily energy demand patterns and gives you a better idea of a particular site's power demands. It can highlight where you could make improvements in your energy efficiency and peak demand, and provides valuable information on the potential sizing of your on-site renewable energy installation(s), as well as battery storage systems.

It is common for larger electricity accounts to have smart meters that record on-site demand in 15- or 30-minute intervals, the information you need to develop a load profile. You can either contact your retailer for the interval data or you might have software systems like SCADA, or a building management system, that provides this information at the click of a button.

Figure 22 shows an example of what a load profile looks like, comparing two 24-hour periods over two sequential days. Because of the dip in demand during daylight hours, particularly around midday, there is likely solar PV on this building. The two days chosen are in winter, so the peaks in the early morning and late afternoon could indicate that heating was turned on. These peaks could potentially be reduced by battery storage. There is also a high base demand, so the question needs to be asked what this organisation could do to bring it down.

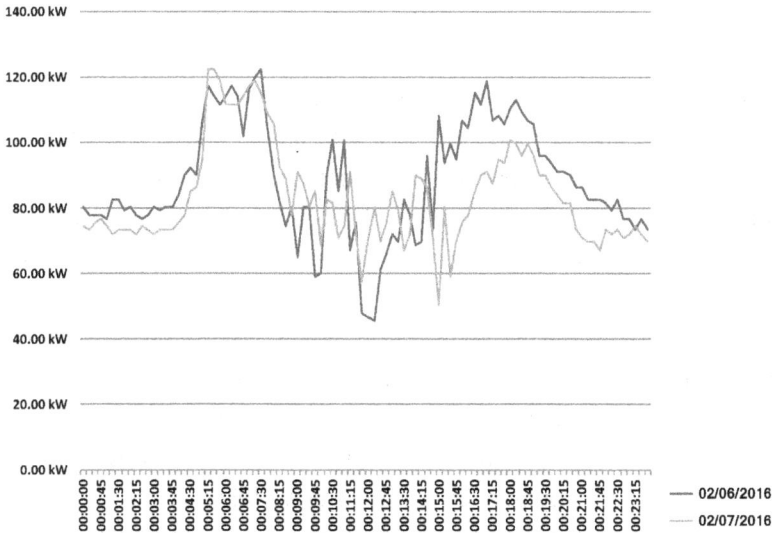

Figure 22: Example of a load profile

Energy audits

Energy audits are a vital tool to uncover site-specific energy efficiency and renewable energy opportunities and determine what their implementation costs and benefits are. Unless you have already improved your energy productivity significantly, you will be surprised how much you can save on operating expenses just by being more efficient with existing resources.

An energy audit not only identifies options to reduce energy usage through upgrading or replacing equipment, installing on-site renewable power generation, and implementing or improving controls; it also identifies and recommends how to enhance energy management practices. The energy audit should also consider the existing operational constraints of each site.

Standards like ISO 50002:2014[165] or AS/NZS 3598:2014[166] cover the three different levels of energy audits available. I recommend reading those standards so that when you engage an auditor, you have an idea of what services to ask for.

An energy audit is conducted by inspecting a particular site to identify where, when, and how your site uses energy and what the building fabric, the site services, their controls, and major energy-using processes are. The site's occupancy, usage patterns, location and climate, environmental conditions, and legal or regulatory requirements are also analysed.

Energy auditors assess your historical and current energy usage over a two-year period, examine your tariffs in detail, and check whether you are being billed correctly. Retail energy rates are compared with market prices. We have saved clients thousands of dollars simply by switching them to a more appropriate tariff or uncovering billing errors.

Another part of an energy audit is the energy balance of a facility which, similar to the one shown in Figure 20, outlines where energy is consumed onsite. It is usually developed based on equipment ratings and hours of operation, site metering profiles, and – where financially viable or where a detailed technical analysis is needed – by sub-metering equipment.

The final report of a typical energy audit shows the site it applies to, the available opportunities, the expected annual cost and electricity savings, and the expected reduction in carbon emissions. It also lists the capital expenditure needed and the financial return of your energy projects.

The energy assessment work of the Green Brewery

An example of an organisation that has undertaken massive amounts of work to analyse its energy performance is the Green Brewery, in Leoben, Austria. When it embarked on its journey to 100%, energy consumption was responsible for 10% of all production costs.

Rather than cut staff costs, they decided to maximise efficiency. It started with an in-depth analysis of energy-using processes by developing energy balance and Sankey diagrams, which show the flow of energy from the source to their end use, and all the energy losses along the way.

Once the brewery team had all the data in front of them, they compared their performance to those of their peers and investigated how to develop more efficient ways to produce beer. It was this in-depth analysis of their energy data that enabled them to pinpoint improvement opportunities.

You, too, after compiling the details about your energy use, will be in a position to investigate which of the renewable energy and energy efficiency opportunities listed in Step 1 might prove a good fit for your operations. The next two chapters dive deeper into the analysis and prioritisation of the opportunities, and how to engage your stakeholders.

Your checklist:

You may be able to delegate or outsource these tasks.

- ☐ Calculate your energy baseline for the fuel sources included in your boundary.
- ☐ If you also have greenhouse gas reduction targets, calculate your carbon footprint baseline.
- ☐ Investigate the expected changes in your organisation and calculate your projected future energy demand.
- ☐ Analyse the load profile of sites that form part of your boundary.
- ☐ Perform energy audits on the highest energy-consuming sites in your boundary.

Chapter 8

Engaging your stakeholders

Some people believe that once a commitment to 100% renewable energy is made, the organisation will automatically support it. This is rarely the case. To inspire people to stand behind your vision and give you the necessary support, you need to engage your organisational stakeholders and get their buy-in.

In most cases, stakeholder engagement is driven from the top. If your CEO or general manager is not fully committed to the 100% goal, the management team and employees will feel less obliged to help achieve the target. Conversely, you may have senior management commitment and a target to transition to 100%, but without the buy-in of your stakeholders, this may not be enough to drive the desired change.

Further, if senior management cannot clearly articulate their organisation's renewable energy commitment, and are unable to link this back to their broader strategic objectives, there is a significant risk of failure.

Stakeholder engagement is about recognising the existing capacities of people to be active participants in the process rather than bystanders, and allowing them to shape the solution. Facility managers, for instance, are in an excellent position to provide support in setting up, maintaining, and monitoring your energy efficiency and renewable energy systems on a day-to-day basis. They are also in direct contact with service providers. It is vital that they buy into your project from the early stages.

Engaging your stakeholders is also about finding and acknowledging the ideas and initiatives that are probably widespread across your organisation, and will help you to formulate your plan. People are your greatest assets, and you will be amazed at what kind of information you uncover just by sitting down with them and asking the right questions.

For instance, in one of my projects, the program manager identified a person working in the water and sewer division who had already investigated one renewable energy option, specifically micro-hydro. Rather than doubling up on the efforts this person had already undertaken, the program manager drew on the findings that had already been made.

He asked what the problems were and why the project did not go ahead. Was it too hard, or were people too busy? Was it not feasible? Was management not supportive? In this particular case, the team was too busy to investigate the opportunity further, so it was decided to include micro-hydro in the list of options to be examined in detail.

Another example is an initial kick-off meeting with a project manager, who was asked how many sites already had solar PV installed or investigated. The project manager was not sure about ones that had been investigated but could identify ones that had been installed.

Talking to the asset managers of the buildings uncovered solar feasibility studies that had already been undertaken at multiple sites. Rather than wasting time and money duplicating the information, it was already there, and just needed to be incorporated into the plan development. This situation can be particularly commonplace in large diverse organisations, where studies are often commissioned by one part of the organisation, and results are not necessarily broadly communicated.

Some multinational companies struggle to engage local employees. Staff far removed from the head office can sometimes feel disengaged from a global plan to move to 100% renewable energy. It can also work the other way round. Sustainability initiatives may be developed locally that the head office will pick up and amplify with additional support.

An example of where stakeholder engagement was driven from a local subsidiary is the Green Brewery in Austria which set itself a goal to become 100% renewable in 2012. The closer it moved towards achieving the goal, the more attention it received from its holding company, Heineken. Heineken soon put its full weight behind the brewery and helped fund some initiatives to make sure the brewery could achieve its goal and stand as an example of sustainability leadership for the rest of the group.

Good stakeholder engagement harnesses the ideas and knowledge of people, identifies opportunities that would otherwise be missed, and overcomes barriers. Other benefits are that it can break up silo thinking, give you a better understanding of on-the-ground issues, and increase your capacity to innovate. It also allows you to share the responsibility and problem solving of the development and delivery of the plan.

How can you engage your organisational stakeholders to ensure your 100% renewable energy project will be a success? The model I follow when helping clients with stakeholder engagement is a three-phase process.

The three-phase process to engage stakeholders

First, list all the stakeholders of the 100% renewable energy program. Keep in mind that you may miss some stakeholders first time around. Next, analyse whether they have a high or low level of interest and influence and map them on a matrix. In the third step, develop a stakeholder plan and use this to engage the various stakeholder groups.

Figure 23: Three-phase process for stakeholder engagement

Phase 1: Identify your stakeholders

Stakeholders in your 100% renewable energy program are people, groups, or organisations who:

- Are interested in the program
- Need to provide input
- Are (or potentially are) affected by the program
- Will be responsible for the implementation
- Have the ability to influence outcomes.

Examples of stakeholder groups are your employees/co-workers, local community, shareholders, suppliers, customers, other local businesses, and the media. I recommend brainstorming methods to capture as many stakeholders as possible in the first go. You can always do a second pass to filter them later.

> An effective stakeholder strategy includes making sure you consider individuals who oppose your 100% renewables goal. Involving them enables you to take their viewpoint and knowledge into account and identify barriers to overcome.

Start with identifying your internal stakeholders. The best way to find these people is to analyse the organisational chart.

- What individuals, or teams and divisions, have a stake or an interest in your 100% renewables strategy?
- Who needs to make relevant decisions?
- Who will impact your strategy and solutions?

Managers are often influenced by recommendations of their key operational staff, and this is not always evident by simply viewing an organisational hierarchy chart. You may need to delve deeper than these hierarchies.

- Who is critical to the delivery of the projects coming out of your strategy?
- What operational areas will likely be affected by your 100% goal?

- Who can contribute resources, financially and in the form of personnel?
- What are acceptable investment hurdle rates?
- Who can slow or stop the project if they are not engaged?

Typical examples of people who need to be involved are the executive sponsor, the chief financial officer, asset owners, the sustainability or energy team, the 100% program manager, facility or operations managers, procurement, the marketing department, and potentially the fleet and waste manager.

> Bring influential employees – and not just executive team members – into the planning process. Not only can they contribute meaningfully to strategy but they also ensure the organisation successfully engages with your 100% goal.

Once you have identified your internal stakeholders, you can turn your attention to groups outside your organisation, like the community, neighbours, industry or regional forums, and suppliers.

- What are the existing businesses in your community?
- Who can help – or hinder – your program?
- What is the local attitude towards energy efficiency and renewables?
- What are your local, state, and federal governments' approaches to energy and sustainability, and what do their support programs look like?
- How does this fit with what you want to achieve?
- Which energy efficiency and renewable energy providers service your area?
- Who is your energy retailer and network operator and what will be their attitude towards your goal?
- What renewable projects or initiatives are already being undertaken that can be leveraged?
- Are there any renewable energy-buying groups you would like to engage?

- Are there any community-based renewable energy projects or industry/regional forums you should establish contact with?

Once these questions are addressed, you will have a list of stakeholders relevant to your 100% renewable energy program. Not all people you identify in this phase will need to be engaged from the early stages of plan development. Some will only need to be involved later, for instance, when you start to implement projects. The timing of the engagement will be determined in the subsequent phase, 'Engage'.

Phase 2: Analyse your stakeholders

In the second phase, you analyse your stakeholders by asking questions like:

- What are their expectations and interests?
- What is their level of influence?
- How easily will they understand and buy into the overall vision?
- Will there be resistance?
- What stake do they have in the project?
- How will they be impacted?
- How influential are they?
- Would they create a problem if their views and concerns were not taken seriously?
- What does the existing relationship look like?

To avoid overcomplicating things, a simple matrix can be used to analyse organisational stakeholders' needs. On one axis, plot their level of influence and give it one of two values, low or high. (See Figure 24 below.) On the other axis, plot their level of interest, also as either low or high.

Then, go through the stakeholder list one by one – it is often beneficial to do this with the program manager – and plot each stakeholder in the most appropriate quadrant. Assigning your stakeholders to the different quadrants allows you to determine the best way to engage each group, which is done in phase 3.

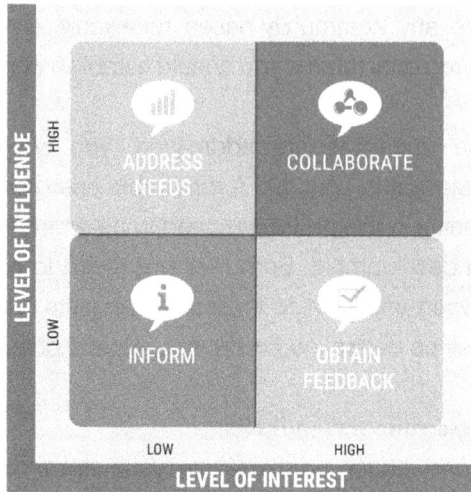

Figure 24: Phase 2 – Use a matrix to analyse stakeholders

Phase 3: Engage your stakeholders

Not all stakeholders need to be involved in the same way or to the same extent. Some people, organisations, or the public may only have a low level of interest and degree of influence in your 100% project; therefore, simply informing them of your goal and your progress towards it should be enough to ensure their continued support.

These stakeholders are best engaged with communication methods that reach large numbers in a simple and effective way. Examples of this are updates on your website, your annual reports, published fact sheets, and newsletters. Once you implement some of the projects, you can even have an open day or site tours to demonstrate your achievements.

People or organisations possessing a low degree of interest but a high level of influence must have their needs addressed, or else you run the risk that your project will be slowed down, even halted. The best way to engage these groups is to have one-on-one meetings, which gives the interaction a personal dimension and assures them their views have been heard and understood. One-on-one engagement also helps to bring people on board, because you can use your influencing skills to show them how their cooperation is necessary to reach the common goal.

This group also benefits when they participate in workshops around your 100% renewable energy plan development, which are ideal to actively collaborate on the pathway development. This is especially the case if you need to discuss complex issues and competing options. Having this group in the workshop generates ideas and builds ownership.

> Workshops must be facilitated to obtain feedback and hear the views of the various participants in open and honest discussions. An experienced facilitator ensures that all participants are equal in contributing to the process, as opposed to permitting a few to dominate the debate.

People or organisations possessing a high level of interest but a low degree of influence also need to be consulted so that you can obtain their feedback. The methods for this group are similar to the previous group – you engage them either through meetings, workshops, your web page and social media, or you survey them.

You can survey your employees, neighbours, or – if you are a municipality – residents. An example of where an organisation conducted an extensive stakeholder consultation through a survey was Lismore City Council in Australia. The council asked residents for input into its strategic vision. The feedback was that residents wanted Lismore to be a model of sustainability, which resulted in the commitment to 100% renewable energy.

Another example is Oxford County, in Canada, which developed a 100% renewable energy plan[167] and used a feedback form on its web page[168] to get residents' views on the plan so it could continually improve it. In Byron Bay, Australia, the community is ahead of the council, developing plans to transition to zero emissions which may, in turn, force the council to keep up.

People or organisations with a high level of interest and a high degree of influence will, without a doubt, be your most important stakeholder group. People who fall into this group should form part of the program steering committee and shape the strategy and solutions along the whole journey. You need to make sure that you have their absolute support and, if possible, have them reach a consensus on the direction for the 100% renewable energy plan.

Preparing your communication material and running workshops

Once you know who your stakeholders are and how you will engage them, you need to plan the timing and the logistics. Consider what communication material is necessary to prepare your stakeholders at each stage of your journey to 100% renewable energy.

- What do they need to know, and when, so they can best contribute to the planning or implementation phases?
- How will you distribute the communication to your stakeholders?
  - Will you send an email, will you display posters, will you put information on your intranet or web page?
  - Will you create a specific email address or phone number so that people can contact you if they need more information about your 100% renewable energy program?
- If you plan to run workshops, how will you make sure you create an atmosphere of trust and collaboration?
- Who will facilitate the workshop and who will capture the outcomes?
- Where will you hold the workshop, given the anticipated number of participants?
- Who will send out the summary of the results to the participants afterwards?
- You also need to think about what resources you need
- Will you have a slide deck, printouts, or flipcharts?
- What activities will you run, and what or who do you need to run them?
- Will you need pens, Post-it notes, a whiteboard, or a screen and projector?

- Does senior management need to be present for the whole workshop, or can you provide time slots for them?

You may benefit from hiring an external consultant for the initial stakeholder engagement. An independent third party will be unbiased and not involved in company politics, and there may be more trust from participants that the best overall outcomes for the organisation will be achieved.

From a personal perspective, this provides you the opportunity to participate in the process rather than having to lead and facilitate it.

To give you an idea of how you might run your workshop(s), let us look at how I typically facilitate mine. I like to create a supportive environment in which participants – even those sceptical towards a 100% renewable energy goal – can explore their experience and those of others, and share their knowledge, skills, ideas, and concerns.

I project a slide deck onto a large screen and I have posters on walls that explain the various energy efficiency and renewable energy options. I present these to get everyone on the same page.

I then ask the participants to work in small groups, to stand up and visit each poster to discuss and engage, with several questions on display to assess the applicability and suitability of the energy option to the organisation. This adds welcome dynamics to the process and often stimulates group discussions about the content.

Individuals are given an opportunity to put their feedback on Post-it notes and affix them to the posters. My team members and I interact with each group to discuss the options, facilitate the discussion, and help to further participants' understanding of each technology option.

If they wish, participants can highlight a technology or a particular aspect they would like to discuss further. I then elaborate on these issues in a group discussion, and a scribe captures the main discussion points from this process.

Next, I allocate 10 dots to each participant to 'invest'. Without discussion, they place their dots where they think their 'investment' will get the best results. We add up the dots and the technology options with the most points are prioritised for further development.

Engaging with stakeholders outside your organisation

If you are going to develop energy projects that affect stakeholders outside of your operational boundary, be sure to provide accurate and timely information about the project's impact and benefits to the stakeholders right from the start.

You may need to engage with your retailer and network provider to understand some of the regulatory barriers and processes in place, under development, or under consideration that may influence your plan. Involving the retailer and network provider also provides you greater insight into expected complexities of your potential renewable energy projects. It is important to understand who the key decision makers are within these organisations and to develop professional and ongoing relationships for mutual benefit.

So that no misinformation is disseminated, it is important to engage with stakeholders in a transparent and timely manner. You can achieve this via local meetings, working groups, and having the stakeholders participate in your meetings.

Listen actively, rather than pushing your agenda without any regard for what really matters to your stakeholders.

You should not assume that community opinion will align with your views about the merits of your energy projects. It is best to engage with neighbouring businesses, local government, and the community early to ascertain the level of support you can expect, and to give them an opportunity to voice their concerns.

One example of where an organisation successfully engaged with its community is the Green Brewery in Austria. It identified a saw mill nearby that had excess heat it did not need. The brewery entered into negotiations, the result of which was a private pipeline that transfers the heat to the brewery for use in its production processes.

Using stakeholder input for your pathway development

The initial stakeholder engagement process will result in a range of influencing information, from acceptable hurdle rates to barriers, to good and bad past experiences, to site-specific factors, to lists of potentially eligible sites for renewable energy developments, and to key people who need to be involved. This information will help you frame each opportunity and identify a series of high-priority options to investigate further.

The best way forward is to take all stakeholder feedback and input into consideration and to evaluate their priorities and competing interests. This is essential to align your pathway to 100% renewable energy with your organisational preferences.

The most important, immediate results of your stakeholder engagement process are that you have identified energy efficiency and renewable energy opportunities which you can now analyse further. The next chapter explores how to do this.

Finally, bear in mind that stakeholder engagement is not a one-off activity; you need to establish a regular cycle of dialogue and feedback with your stakeholders to ensure continued buy-in. This helps you keep abreast of potential risks and opportunities, and stakeholders see how you respond to their views. Done well, stakeholder engagement builds trust, enhances relationships, and enables a smooth transition to 100% renewable energy.

Your checklist:

You may be able to delegate or outsource these tasks.

☐ Develop a list of key stakeholders, inside and outside your organisation.

☐ Analyse the needs of your identified stakeholders.

☐ Engage each identified stakeholder group according to their interest and level of influence in your 100% renewable energy project.

☐ Prepare targeted communication material for your stakeholder group.

☐ Communicate with your stakeholders and collect feedback.

☐ Consider running a workshop with key organisational stakeholders to brief them on the goal and get their feedback for pathway development.

☐ If you are going to develop energy projects that affect stakeholders outside your organisation, start to engage them early in the process.

☐ Select energy efficiency and renewable energy opportunities that match your organisational preferences for further analysis.

Chapter 9

Analysing and prioritising energy opportunities

Having analysed your energy baseline and successfully engaged your organisational stakeholders has given you a good indication of your main focus areas. The next stage is to analyse your opportunities in greater detail to select the most suitable ones for your pathway.

Some of your energy projects will be quick and inexpensive to implement. Some may not deliver large energy savings. Others may require substantial investment and much planning, but will significantly contribute to your target. Some opportunities may provide modest savings, but strategically align with your organisational preferences or the competencies you may want to develop.

Your pathway comprises a mix of these opportunities, but a systematic way to analyse and prioritise them will enable you to improve them further by identifying the best financing and delivery options.

Analysing your opportunities

The work you did at the beginning of Step 2, like analysing your energy consumption and engaging your stakeholders, identified a number of energy efficiency and renewable energy opportunities, which you have selected for further analysis. A typical example is energy consumption reduction, or onsite renewable energy generation, like solar PV.

However, you may have also identified opportunities like biofuels, electric vehicles, staff energy awareness and driver training, or purchasing renewable energy certificates and carbon offsets. To assess how applicable these opportunities are for your sites and circumstances, you can assign them to skilled staff, or consider engaging experts who can perform and provide detailed analyses.

Several terms are commonly used to describe the assessment of energy efficiency and renewable energy opportunities, like energy audits, modelling, or feasibility studies. All of these share a common purpose of proving the technical merit and savings capacity of your opportunities.

If you have not previously undertaken energy audits, you or an experienced third party need to perform detailed site visits to assess the suitability of your locations for energy efficiency improvements and implementing renewable energy opportunities.

During these detailed site visits, you or your consultants should engage with operational staff to determine the scope and boundaries of your opportunities and any barriers or restrictions, such as physical space constraints, planned changes in use, scheduled upgrades, or whether any latent conditions apply, for instance, the presence of hazardous materials, like asbestos. You will likely also perform modelling to see which solutions would be the best fit within the identified boundary.

This entails investigating the capacity of an opportunity to reduce fossil fuels and evaluating different approaches or possible solutions. Your assessment should also look at how the project could be structured or delivered. For renewable energy opportunities (including battery storage), you may also evaluate whether, and how, the energy project can be connected to the grid. Metering and monitoring of your initiatives should also be incorporated into your investigation.

Detailed assessment of energy efficiency opportunities

You will typically evaluate the potential for various technologies, like LED lighting, variable speed drives or more efficient motor systems, hot water system upgrade and controls, chiller and other HVAC upgrades, building management systems, IT system optimisation, and appliance replacement and control strategies to contribute to your target across all your sites.

Detailed assessment of renewable energy opportunities

When investigating renewable energy projects – both behind and in-front-of-the-meter (more details about the differences can be found in Chapter 3) – you need to know what network infrastructure is located close by and whether it has capacity to accommodate your project. Connecting to the grid can be both time consuming and complex, so engage with your local network provider early.

For every location that is suitable for **behind-the-meter solar PV**, solar modelling experts match the solar output as closely as possible to your load profile within the constraints of your facility. If you mainly have night-time demand, for instance, like in a sports stadium or an entertainment venue, then solar PV does not make much sense unless it is combined with a battery storage solution.

Your modelling expert (who could be a solar installer or an independent party) assesses space for panels, taking into account shading, orientation, wind-loading aspects, structures, and other obstacles. They should also consider the optimal design and placement of the panels to ensure space is not wasted for future extra installations if exporting excess electricity becomes financially viable. This also relates to making sure that boardwalks are not excessively wide, especially on space-constrained sites.

If you plan to install **batteries**, the assessment determines whether you have a suitable space that protects them from theft and the weather and complies with relevant standards.

Your opportunities prioritisation may also have identified preferences for **larger renewable energy generation opportunities**, such as solar PV, wind, bioenergy or micro-hydro, for example. For these opportunities, you may undertake a high-level assessment to prioritise those with the greatest likelihood of success.

If you have identified a **bioenergy** opportunity, you will need to assess the quantity, quality, and seasonality of your feedstock, transport needs, final product, and processing needs, as well as identify suitable locations. (Chapter 3 explored a range of mature, pre-commercial, and emerging technologies to process your bioenergy feedstock.) To determine the feasibility, assess the current status, as well as future prospects.

For potential **micro-hydro** locations, such as waterways and water delivery systems, it is necessary to analyse the flow and head pressure of the system over the course of a year or more. A typical rule of thumb for hydro projects is that 50–60% of available power can usually be converted into electricity generation output, providing an indication of the expected savings or revenue from such a project. However, costs for micro-hydro projects tend to be highly site specific.

Wind energy potential depends on a range of factors, most notably wind speed and swept volume of turbine blades. Australia has well-developed wind energy maps, and you can carry out a high-level screening by referring to these resources for your locations or region. Most wind power development in Australia is sited in locations with wind speeds exceeding 7.5 m/s at 100-metre height, along with other favourable factors.

Mid-scale solar PV projects are attracting much attention. If this is an option for you, identify what land you own or could use that satisfies a range of initial screening criteria. These could include whether it is flat or nearly flat, has the right aspect with enough buffer to residents, is not subject to other governance such as vegetation management or protected species plans, is not located on a flood plain, is not already designated for a key project or development, and is near the electricity grid.

You can find out more about renewable energy resources available in Australia through the ARENA-funded Australian Renewable Energy Mapping Infrastructure (AREMI) website, available at www.nationalmap. gov.au/renewables. AREMI contains information on wind, solar, geothermal, hydro, wave, and bioenergy resources (in development) and is a great tool for finding out what renewable energy technologies are most suitable to your particular location.

Locations for your energy projects do not always have to be within your organisational boundary. You may identify sites that are better suited technically and owned by others with whom you can strike a partnership agreement; for example, if solar resources are better inland, a coastal organisation may seek to develop a solar PV project with an inland organisation. There may also be opportunities to cooperate with another region or other buyers, through PPAs (power purchase agreements), for instance.

Detailed assessment of transport energy opportunities

If transport is part of your 100% renewable energy boundary, evaluate the current state of biofuel development in your region, as well as biofuel quality standards. Experts would undertake an assessment of the feasibility of biofuels for your passenger, commercial, or heavy vehicle fleets, as well as the potential to change your fleet to electric vehicles, including issues like range or infrastructure like charge points.

Assessment of your social licence

In addition to assessing technical feasibility, you may also need to consider environmental and social feasibility for some opportunities, like bioenergy projects, or the use of land for renewable energy developments. For projects like these, you must have the support of the community to have the necessary social licence to operate.

It is also important to investigate perceptions of the economic, environmental, and health impacts, how your project will affect local biodiversity, air quality, and visual amenity, or the availability of local jobs. For bioenergy projects, it will also be a question of what the particular feedstock is, and how far it is being transported. (You can find more information on how to successfully engage your community in Chapter 8, 'Engaging your stakeholders'.)

Costs and benefits of your feasible opportunities

It is important to develop an initial cost-benefit analysis for your opportunities, so that in the next piece of work you can look at how to deliver and finance them. The costs for renewable and energy efficiency technologies depend on their technical maturity, level of commercialisation, market availability, and the current support that comes in the form of government assistance.

Technology costs vary, depending on the region you are in, or even the country. There are global downward cost drivers, like falling cost curves for renewables, and dwindling government support for fossil fuels, which is making renewable energy increasingly cheaper.

> While, generally, costs are falling, your particular project's costs (and benefits) are dependent on your site. I recommend you undertake site specific financial analyses to find out whether your solutions are cost effective.

You can determine the value of savings for your *energy efficiency projects* by multiplying the expected energy savings by your tariff. The exact amount of savings depends on the specific opportunity, the time-of-use of savings, how your retail and network rates are structured, and on potential government incentives, such as feed-in tariffs and energy efficiency white certificates. (White certificates are documents certifying that a certain reduction of energy consumption has been attained.)

You should also be mindful about future changes in energy prices, the depreciation of your assets' values, and other considerations related to the time-cost of money. I suggest using a conservative approach to forecasting when projecting savings, like a low escalation rate for energy prices. When in doubt, invest in expert advice to help you obtain forecasts. (You can read more about financial appraisal techniques in Chapter 13.)

If a particular site is on a time-of-use tariff, for instance, and you implement shutdown procedures at night or on weekends, when your business does not operate, then the project value is the energy savings multiplied by off-peak energy rates. If your site is on a tariff with peak demand charges and your energy efficiency project can reduce power consumption during peak times, then the value of the savings will be higher, as you will apply peak time energy tariffs, and possibly peak demand charges.

Let's take the example of a high bay lighting to LED upgrade in a warehouse. The warehouse operates between 7:00 am and 10:00 pm on weekdays only, which matches peak and shoulder periods. The lighting upgrade will save 300,000 kWh per year.

To evaluate the savings of the project, multiply that 300,000 kWh by a blend of the peak and shoulder rates, say, 15 cents/kWh, which results in savings of $45,000 per year. However, the project also reduces peak demand by 85 kVA every month, for which the business currently pays, say, $10/kVA. This results in additional savings of $10,200 every year. The total value of energy savings for this project comes to $55,200 per year. Because of the longer life of lighting technologies such as LED and induction, you will also benefit from maintenance savings through fewer lamp changes.

Apart from electricity savings, you might gain additional benefits from your energy project, like reduced maintenance expenditure which can be quantified and reflected in your cost-benefit analysis as well.

The value of your *renewable energy projects* depends on whether they are located in front or behind your meter. If your renewable energy project is *behind* the meter and all the generated energy is consumed on-site, your project will offset the price you would usually pay for electricity at the time of solar power generation. For a commercial business, this often means that peak or shoulder period bundled energy rates (per kWh) are avoided, and the site's peak demand is often reduced.

However, if your renewable energy project is located *in front* of your meter, you will compete with other generators connected to the network. You will be paid wholesale electricity prices for your produced energy, which translates into getting less money for your power generation. On the other hand, you may be able to sell renewable energy certificates for your generated power and receive money for the surplus electricity you put into the grid.

The value of your *transport energy projects* depends on the price you currently pay, and expect to pay in future, for your vehicles and fuel.

Future costs of emerging technologies and fossil fuel prices

Several emerging technologies may play a role in your energy plan over the long term, so it pays to research when they are expected to become more cost effective, as this can affect the staging of your actions towards your 100% renewable energy goal. With falling cost curves for most renewable energy technologies, it may only be a few years before technologies that seem out of reach now can become part of your strategy.

To get an idea of future costs, research reputable sources on the predicted costs of specific technologies, how their availability and cost will change in the next decade, and whether these changes will make them attractive to you. You should also be able to target an exact price point for the technology that would make it financially feasible for your organisation.

In my experience, cost curves always appear more conservative than actual developments, meaning that technology innovations, improvements in manufacturing technology, economies of scale, increased uptake, and the removal of regulatory barriers to renewable energy are bringing costs down faster than most models predict.

For instance, when developing a renewable energy master plan for a client in 2014, we investigated the expected costs of technologies at a certain point in the future. One year later, we revisited those costs, only to find the price had dropped much more rapidly than had been predicted in the models.

You can also expect fossil fuel costs to become much more volatile, which will also make your renewable energy projects comparatively cheaper.

Prioritise your opportunities

To prioritise all the identified opportunities, engage with decision makers to gain consensus. You can approach this in several different ways, but here are two approaches that work well.

The first is to have a series of one-on-one meetings with key asset owners and decision makers and subsequently follow up with a senior management presentation and discussion. The purpose of the presentation is to inform management about the work and get the prioritised opportunities accepted.

Alternatively, you can run a workshop with both key asset owners and decision makers, in which you explain the opportunities with the intent to bring participants on board, get their suggestions, and improve or refine the opportunities. You then seek sign-off on the prioritised opportunities.

For both the first and second methods, you can present your energy baseline, the fossil fuel reduction from each energy opportunity, their costs and benefits, the expected social climate, and the feedback you received from your stakeholder consultation process and the existing resource capacity (both human and financial) to deliver various projects.

If you run a prioritisation workshop, participants can consider a range of inputs, which will be used to prioritise your energy opportunities. Examples of such inputs are organisational preferences like:

- Whether your organisation would rather own and control energy assets
- Whether you would prefer to lease them, or simply purchase the generated output
- The importance of long-term energy price stability
- Whether energy projects need to be made visible, i.e. can be seen from the street
- The location of your renewable energy projects – should they be close by or could they be anywhere so long as they satisfy your financial criteria?
- The acceptability of purchasing renewable energy certificates
- Generating your own renewable energy.

Other inputs that matter to your organisation are benefits like the annual renewable energy production or energy use reduction, ease of implementation, costs versus the budget you have available, the financial delivery options you prefer, the co-benefits of your opportunities (such as reduced maintenance), internal timing requirements or constraints, and whether management can agree on what energy projects are to be implemented.

If you like, you can use the 'investment' dots method I described in Chapter 8 and tally up the number at the end to see which opportunities attract the most points. These opportunities will be developed further and form part of your strategy to 100% renewable energy to be approved by management.

Rather than running another workshop in which you present possible delivery and financing methods, you could roll this discussion into the opportunity prioritisation workshop and get consensus on which of these is preferred.

Immediate next steps are to think about ways your projects can be delivered and financed. Nowadays, there are many options available, which are discussed in greater detail in the next chapter.

Your checklist:

You may be able to delegate or outsource these tasks.

- ☐ Perform a detailed assessment of the opportunities that came out of your stakeholder engagement process.
- ☐ Perform detailed site visits to determine the scope of your opportunities and any barriers that need to be managed.
- ☐ Engage with energy efficiency and renewable energy providers, as needed.
- ☐ Model the capacity for energy efficiency or renewable energy initiatives to reduce your fossil fuel energy consumption.

☐ Investigate the technical feasibility of the preferred energy opportunities.

☐ Investigate the social licence for the preferred energy opportunities.

☐ Perform a cost-benefit analysis of your energy opportunities.

☐ Look up the cost curves for renewable energy technologies that might currently be out of reach and determine if, or at what point in the future, they might contribute to your target.

☐ Shortlist your energy opportunities by further engaging key organisational stakeholders.

Chapter 10

Delivering and financing your opportunities

At this stage, you possess a comprehensive understanding of your organisation's energy profile, have engaged key stakeholders and got their support, and have a set of prioritised preliminary energy efficiency and renewable energy business cases to achieve the 100% goal. Now let's look at the various options to deliver and finance your opportunities.

Historically, renewable energy assets have been owned and funded by large private companies or governments. This has changed considerably and is one of the biggest innovations in the energy space. Increasingly, renewable energy generators are owned and financed by businesses, energy service companies, or by the community.

When you implement a renewable energy project there are several roles involved in setting up and running your project:

- Host
- Owner
- Developer
- Financier
- Operator
- Retailer

- Network operator
- Energy buyer.

Your organisation can decide to take up most of these roles, or only one or two and outsource the rest, depending on your circumstances. For instance, you could host a behind-the-meter solar PV installation owned and financed by you but installed and maintained by a solar company.

You could host a mid-scale solar PV in front of the meter (see Chapter 3), owned and funded by the community, whilst your role would be to buy the energy. You could bulk-purchase power from a wind farm that is geographically located far from your site, owned and operated by a wind farm developer, and financed by a third party.

Tying into the delivery option is the financing of your energy project. Previously, there were few options to invest in energy efficiency or renewables – to buy and own the equipment, or finance it with a loan.

In recent years, innovative, more flexible financing mechanisms have entered the market. You can now decide on how to structure the cash flow, what party will own the asset and the associated risks, and how the asset will be treated on the balance sheet.

On one end of the spectrum, you can self-fund all the actions, and own and operate the new assets. IKEA, for instance, committed to own and operate wind turbines in 10 countries it does business in.[169] For example, in early 2016, it bought a 42 MW wind farm in Finland.[170]

On the other end of the spectrum, you can have no upfront investment, outsource the ownership, operation, and maintenance, along with the risk the project might not perform as expected, and only pay monthly instalments from the operational budget. This is often described as 'renewable energy as a service'. This approach is similar to the current supply paradigm for many end users, in that you only pay for the energy consumed and merely change the source of the bill.

To get the best possible outcome under your current circumstances, consult divisions like finance, tax, and legal. Examples of considerations are your organisation's legal structure, size, current debt levels and credit rating, time in business, sources of income, profitability, cash flow, the energy project's value, your risk appetite, general market conditions, existing support from the government, and the type of project being undertaken.

The following sections delineate the most common energy efficiency and renewables investment options. You can find further information and case studies on some of them in the Energy Efficiency and Renewables Finance Guide,[171] as well as in the Solar Finance Guide[172] published by the NSW Government. You can also find a summary table, giving you a good oversight of most details at one glance, as well as other resources, on www.barbaraalbert.com.au.

Self-funded from your capital budget/balance sheet

The energy projects can be financed with an organisation's own funds from the capital budget. Because capital expenditures are major purchases, and their costs can only be recovered over time, companies ordinarily plan for these purchases separately from preparing an operational budget.

The advantages include no ongoing contractual obligations and that your business can own and depreciate the equipment. You carry all finance and performance risks and are responsible for the maintenance, unless you outsource this to another party.

Self-funding can be the most cost-effective option to implement energy projects, as all benefits flow directly to your organisation. The overall financial outcomes may be greater, as the annual rewards are not shared with a third party.

Just like with any other capital project, the energy projects will likely need to meet your organisation's minimum acceptable rate of return on capital (hurdle rate) and compete with other business priorities. The energy asset will be on the balance sheet, which is a disadvantage for private sector companies, as it will be reflected as a liability or debt. This can have an adverse effect on financial ratios, like profitability margins, current ratio, and leverage.

Councils, on the other hand, may face political challenges when incurring debt through the capital budget process, particularly if they have to raise taxpayers' rates or introduce special levies as a consequence. Councils also do not have to pay tax and thus do not enjoy any benefit from depreciation schedules.

Self-funded from a revolving energy fund

To avoid having to compete for funds with other projects and business priorities, organisations sometimes set up a revolving energy fund (REF), which is also known as a green revolving fund, revolving loan fund or sustainability revolving fund.

An REF is an internal fund that provides financing to implement energy projects that provide cost savings. The savings are tracked and used to replenish the fund for the next round of investments, thus establishing a sustainable funding cycle, which cuts operating costs.

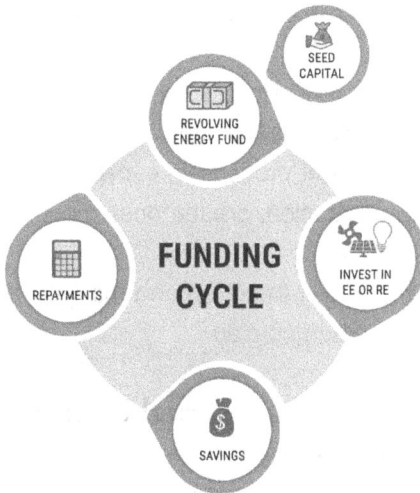

Figure 25: The payment flows of revolving energy funds

To see whether the energy project can be paid for using the anticipated savings it will generate, it is necessary to develop a cash flow from the day the energy project incurs cost to a specified date in the future.

The seed capital for the REF can either come from an annual operating budget or the capital budget. There may be a one-time infusion of capital or multiple injections over time to scale the fund gradually. It is essential that the portfolio performance of all the energy projects and the cash infusions be forecast, to see whether the fund will grow or deplete over time.

The advantages of an REF are that it cements your commitment to the 100% goal and provides a tangible vehicle to achieve it. On the other hand, there might exist internal hurdles in setting up the fund. I have found that another barrier for REFs is that organisations are not aware of this financing vehicle and the manner in which it operates. Moreover, in the case of local governments, a resolution might be required.

As with the other self-funding option, you carry all finance and performance risks and are responsible for maintenance, unless this is contracted out.

In my many years of work in this sector, I have found that the popularity of REFs comes and goes, but organisations that ensure stakeholder buy-in and set up stringent processes, especially around governance, measurement, and verification, are the ones most successful with this strategy.

You can find a checklist to set up and maintain a revolving energy fund on www.barbaraalbert.com.au.

Self-funded through an internal price on carbon

When the Paris Climate Agreement was signed by 175 countries on 22 April 2016, the UN Global Compact[173] called on companies to take a lead in transitioning to a low-carbon economy by setting an internal carbon price at a minimum of US$100 per metric tonne.[174]

Based on work with about 70 companies that have already internalised a price on carbon, the UN Global Compact believes that $100 is the minimum amount needed to spur innovation, unlock investment, and shift market signals in line with the 1.5 to 2-degree Celsius pathway.

As of September 2016, according to a report by CDP,[175] more than 1200 companies currently have an internal price on their carbon emissions or intend to do so in the next two years. For comparison, in 2014, only 150 companies were pricing carbon.

An internal price on carbon is voluntarily set, and means that companies put a price of their choosing on the carbon emissions they are responsible for. Around a third of the organisations in the CDP report disclosed the price they use, which ranges from less than USD1 to more than USD800.

The funds raised can be used to finance energy efficiency and renewable energy projects. An internal price on carbon makes your energy projects comparatively cheaper, which means the business cases are approved more easily and that payback periods are shorter.

A company that has been the forerunner with an internal price on carbon is Microsoft,[176] which has achieved carbon-neutral status with this strategy. The carbon fee is paid by divisions across the business, and funds internal efficiency upgrades, power purchase agreements, green power instruments, carbon offset projects, and supply chain engagement. Other examples of organisations with a price on carbon are Colgate-Palmolive, BP, General Motors, Insurance Australia Group, Australia and New Zealand Banking Group, and Qantas.

Loan-funded

A loan-funded model means that a lender provides capital to your organisation, to be repaid by a given date, typically at a pre-determined interest rate that moves in line with a reference lending rate. You make regular repayments to the lender to cover interest costs.

If the monthly financing is less than your asset's monthly energy savings, this might prove a good option for you. The advantages are no or reduced up-front costs and the interest and depreciation of the equipment may be tax deductible. The disadvantages are that your organisation bears the economic and technical risk should the equipment become unusable, and the loan is on the balance sheet.

> Rather than secure a loan to cover the entire project, you could choose to borrow only enough to cover a portion of the total project cost, with the remainder coming from balance sheet financing.

Operating and capital leases

With an operating lease, a supplier installs the equipment and your organisation makes monthly repayments on the system for a period of time, commonly five to 10 years. The repayments can be a flat monthly rate, or increase during the contract, which is often linked to CPI increases. A capital lease is similar to an operating lease, except that at the end of the lease, the equipment ownership is transferred to your organisation on payment of an agreed amount.

With an operating lease, the financier owns the equipment and your organisation obtains the sole right to use it. Usually, the supplier is responsible for the maintenance of the system during the leasing period, although that is not always the case in real-world applications. At the end of the lease, you have the option to return the equipment, negotiate its purchase, or continue to rent it. In most cases, the host purchases the equipment after a few years.

Leases allow spreading out of the cost of an investment, but it means that repayments with interest are incurred. This makes the equipment more expensive than had it been paid for up front. Operating leases are more suitable for capital-intensive projects and where costs are mainly for physical assets.

They are not as beneficial for less expensive equipment, such as lighting upgrades, or when a significant portion of the expenses are for installation and associated services. They are also less suitable when the equipment is difficult to remove or reuse. The advantages are that there are no or reduced up-front costs and that leasing costs are tax deductible. Leasing can be a financially attractive option if your company has limited cash flow.

A disadvantage is that you have extra costs of interest payments without being able to depreciate the asset. If you are a local Government, however, leasing or outsourcing depreciating assets has its merits, as there are no tax advantages for depreciation.

When assessing the economic benefits of leasing, investment indicators such as rate of return or payback period are not relevant, as there is no upfront capital investment. This is why I recommend calculating the net present value of a straight purchase and comparing that to the net present value of a leasing option. (Chapter 13 contains more information about financial appraisal methods.)

Operating leases were a popular strategy to keep energy projects off the balance sheet and remain cash flow positive, or neutral, for the first few years, after which the full financial benefits would be realised. However, the new International Financial Reporting Standards (IFRS) 16 Leases standard eliminates the classification of leases as either operating or finance leases for a lessee. Instead, all leases are treated similarly to finance leases. What this means is that lessees need to account for leases on the balance sheet from 1 January 2019, unless the asset term is less than 12 months and the asset is a low value.

On-bill financing

It may be worthwhile to check with your electricity retailer to see whether it offers on-bill financing. Under this funding option, the energy retailer installs the equipment and you pay it back, over time, through a 'repayment' charge on the energy bill. The repayment liability appears on the organisation's balance sheet until it is paid off. Once all payments are made, the title for the equipment transfers to your organisation.

The advantages are no or reduced upfront costs and the interest component of the repayments may be tax deductible. The payment through the utility bill reduces the risk of default, which can lower financing costs. If you default on the debt repayment, you risk your energy supply being cut. Also, if energy savings are not guaranteed, you bear the technical risk of the energy project not performing as expected. Additionally, the repayment period is usually short when compared to leases or asset financing, which can make managing cash flows more challenging.

On-site power purchase agreements

For most organisations, power generation is not their core business activity and many prefer to task a specialised company to design, build, own, and operate the equipment. Onsite PPAs are suitable if you have land or space but do not want any upfront cost or involvement with the renewable energy project, other than to obtain power from it.

The PPA provider leases space from you for a set contractual period of time and retains ownership of the system. They sell the generated energy to you, and can also sell you the associated renewable energy certificates (RECs). Dependent on how you have expressed your goal (see Chapter 2), you may need to purchase and retire these, at the latest, from your target year onwards.

PPAs have a typical duration of 10 to 15 years with a minimum of around five years. It depends on the PPA contract as to whether the ownership of the equipment will be transferred to your organisation during the contract time, or whether the PPA contract includes an option to renew.

There are generally no upfront costs for a PPA, and they can be cash flow positive from the day the system is commissioned. They also allow for predictable energy pricing, which is why many of my customers find them attractive. Despite these obvious advantages, I recommend making an informed decision by analysing the economics of both the ownership and the PPA business case, using their net present value.

The more electricity your renewable energy installation can produce, the better your net financial position. Aim for a lower price than you currently pay for grid-supplied electricity.

PPAs are split into 'consumption' and 'generation' PPAs. A consumption-based PPA is better from a risk perspective as you only pay for the energy you *consume* from the system installed. This ensures the provider designs the system correctly to maximise consumption and realise returns.

A generation-based PPA often looks a more attractive offer, but you will be charged for all energy produced by the installed systems whether you use it or not. If, for instance, your business does not operate at weekends or during holiday periods, or if you undertake energy efficiency measures, you would still pay for the energy generated and exported to the grid.

> With generation-based PPAs, beware of sizing the system correctly. If you oversize and don't use all the electricity produced, you must still pay for all the energy you agreed to purchase.

Some of my clients initially find it difficult to understand the differences between leases and PPAs. I explain that with PPAs, they commit to buy a certain amount of *energy* per month, whereas with a lease agreement, they commit to spend a certain amount of *money* per month. PPA providers also take care of all the technical and performance risks over the life of the PPA contract, as opposed to lease or purchase options. Many also have optional maintenance and monitoring packages available.

Off-site power purchase agreements

Similar to the business case for on-site PPAs, to become a renewable energy owner and operator requires a commitment to develop suitable in-house expertise that may not be reflected in your business plan or corporate vision. You may also not possess the required space to host your own renewable energy installation. This is when off-site PPAs come into play.

The renewables development could be situated where enough energy potential exists from a renewable source at a low-cost location. The developer is responsible for the design, construction, ownership, and operation of the equipment, and you agree to purchase the renewable energy certificates and/or the energy output from the power plant.

Historically, the buyers of electricity generated from large-scale renewable energy projects have been utilities. Increasingly, it is organisations that want to meet their sustainability targets and purchase the power directly from developers.

The price of green tariffs or RECs that a retailer purchases on your behalf can fluctuate, introducing uncertainty. Striking a PPA deal hedges against those price fluctuations, not only of the RECs but also generally of fossil fuel-based energy.

Despite their advantages, PPAs are still in the early stages of acceptance. Most organisations – even when they spend millions of dollars on electricity per year – are used to buying electricity from their retailer, or procuring power on a short-term basis in the spot (short-term energy) market. It is an entirely different decision to enter into a 10-, 15-, or 20-year PPA.

> PPAs are long-term contracts and, as such, can take time to obtain internal approval, so remember to take organisational planning time into account.

Having commercial purchasers, or 'off-takers', that enter into long-term contracts can make or break a new renewable energy development. You help more renewable energy projects come online when you enter into PPAs.

Purchasing renewable energy from a power plant located off-site (see Chapter 3) is considered to be equivalent to the direct generation of renewable power generation onsite. On that basis, the voluntary cancellation (retirement) of RECs is treated as a zero emissions electricity source in a greenhouse gas inventory.

The closer the renewable energy project to your location, the fewer network losses will be incurred, and from a social perspective, the more the local area will benefit from your investment. I once visited a wind farm in Austria, one of a handful worldwide that allow you to climb to the top to a viewing platform. It provided a fantastic view of all the other nearby wind turbines and the land underneath, which was used for farming. Every wind turbine needs to be accessible and we could see all the roads that led to each windmill.

Our guide mentioned how happy the local government was with the developments, as these roads were now heavily used by the locals and tourists to walk their dogs or ride their bikes. However, it is not only the additional infrastructure that gets created through such projects but also local jobs, like maintenance work.

One example of a company that uses offsite PPAs to meet their 100% renewable energy commitments is McDonald's UK, for instance. The organisation signed 20-year PPAs with renewable energy developments[177] and claims anticipated savings in the range of £75 million to £100 million.

Other companies that buy directly from utility-scale renewable energy projects include Apple, Facebook, Google, Microsoft, Walmart, Mars, IBM, Amazon, Procter & Gamble, General Motors, and Dow Chemical.[178]

Depending on the regulation of the energy market, a company can also enter into virtual PPAs (VPPAs). VPPAs are also known as corporate or synthetic PPAs. Under a VPPA, you agree to pay a renewable energy project developer a fixed rate per kWh over the contracted period of time whilst you continue to purchase electricity from your regular retailer.

The advantage is there is no change to the existing agreement with your retailer. In exchange, the developer generates and sells power on the open market, turning the proceeds over to you. If the rate agreed on by your organisation and the developer is below the market price, you make a profit.

An example of a company that sources energy through a VPPA is Salesforce, which has entered into 12-year agreements for 64 MWs of power with two wind farms.[179] The developer puts the energy onto the open market, which Salesforce committed to purchase, regardless of the cost of sale.

Community ownership

The most common finance structures for community ownership are either PPAs or community loan arrangements. With a PPA arrangement, the renewable energy project is developed and owned by a cooperative or company made up of community members. Your organisation can be one of those community investors.

The upfront costs for the renewable energy system come from community group investors who may include municipalities, local NGOs, charities, sustainability groups, 'mum and dad' investors, and businesses. The upfront cost and shareholder profit is recouped via ongoing PPA payments from organisations that participate as hosts and buyers/users of the renewable energy. Such a finance arrangement is exemplified by Repower Shoalhaven, in Australia.[180]

An example of a community loan arrangement is Lismore City Council's 'Farming the Sun' project.[181] Lismore Community Solar raised funds through a private share offer from around 40, mostly local community investors to lend to the council, who in turn has pledged to build and operate two 100 kW solar plants on council land, and repay the loan over a seven-year term at a fixed interest rate of 5.5%.

The council will begin to make its money back in 10 years. With this model, the council pays a slight premium on the loan interest, but the motivation behind the project is that it is a community energy partnership project, that builds capacity in the local community.

Typically, what makes these projects so attractive for organisations is the strong community support and associated media coverage. Community renewable energy projects have financial benefits but also add local renewable energy capacity to the grid, ensure money from renewable energy investments remains in the community, create local jobs, provide a vehicle for ethical investments, strengthen community networks, build energy capacity in the community, and provide the community with a landmark renewable energy project that could attract further investments. They are also a way for residents of apartment buildings to invest in and benefit from solar.

In Denmark, renewable energy developers must, by law, offer 50% of the shareholding in the project to residents living within two kilometres of the project. This has led to Denmark becoming a world leader in renewable energy. Community ownership is also popular in Germany, the UK, and the US. The first community-owned renewable energy scheme in Australia was the Hepburn wind farm in Victoria, owned by 2300 members, which became operational in 2011.

Energy service agreements (ESAs) and energy performance contracts (EPCs)

An ESA is an established model previously based on co- or tri-generation and energy efficiency projects, which in recent years has extended to include renewable power generation. Rather than being a stand-alone finance option, an ESA provides end-to-end delivery and ongoing operation of energy projects. Financing can be arranged using funds from your budget, a loan, a lease, or be organised by the ESA provider itself.

ESAs have been in common use in the US for a few years and are becoming more common in Australia. An ESA provider designs, constructs, owns, and operates the equipment. You pay fees to cover the operation and maintenance costs, the energy expense, and the repayments of the capital and implementation costs. You can usually purchase the equipment at the end of the ESA.

The advantages are no or reduced upfront costs and that the ESA may be treated as off the balance sheet. The payments are likely tax deductible and the implementation and operating risks are transferred to the ESA provider. The ESA is also incentivised to maximise the energy savings.

Your organisation will end up with two energy bills: one for the balance of power still required from the grid, and one from the ESA vendor for energy you would have needed to purchase. You will be better off financially as long as the money you pay to the ESA provider is less than you used to pay your retailer for the same energy.

One disadvantage is that the costs can be higher than with other finance options, as you have to pay to transfer the risk to the ESA.

Other financing options include energy performance contracts (EPCs). EPCs allow funding energy upgrades from cost reductions. These contracts are a means to deliver infrastructure improvements to organisations that lack energy engineering skills, time, capital funding, understanding of risk, or technology information.

Under an EPC, an energy service company (ESCO) is engaged to improve energy efficiency or deliver a renewable energy project. The ESCO uses the stream of income from the cost savings, or the renewable energy produced, to repay the costs of the project, including the costs of the investment. Essentially, the ESCO does not receive its payment unless the project delivers energy savings as expected.

The ESCO is responsible to implement and usually maintain the energy savings program, and the management fee is not paid if the guaranteed savings are not achieved. In this way, the technical and, in some cases, financial risks of energy savings programs are transferred from the customer to the ESCO.

Grant-funded

In certain cases, and if you are lucky, you might be eligible for grant funding or other subsidies from the local, state or federal government. The government might fund your whole project, it might co-fund, or it might provide you with specialised expertise.

Grant funding opportunities can arise quickly and often give only a few weeks to receive submissions. You might consider having a project or two 'shovel ready', so you can act quickly should a grant arise.

Access to grant funding and the specific eligibility requirements continually change. Having individuals within your organisation who are aware of these changes and subsequent opportunities can provide real value. Alternatively, you can ask your trusted advisers for help.

Australian organisations can visit https://www.business.gov.au/ assistance to find out more about what is available. Consider contacting institutions like the Clean Energy Finance Corporation[182] (CEFC) or the Australian Renewable Energy Agency (ARENA) as well.[183]

Your checklist:

You may be able to delegate or outsource these tasks.

☐ Engage your key organisational stakeholders to find out which delivery and financing options they prefer for each of your broad categories of energy efficiency and renewable energy opportunities.

☐ Engage with service providers that offer suitable funding methods.

☐ Find out whether grant funding or other financial assistance is available for your energy opportunities.

Chapter 11

Building your pathway to 100% renewable energy

Making the commitment to 100% renewable energy can feel like an enormous target. You might wonder how you can change your whole energy supply from conventional fossil fuel sources to one based on renewable energy.

However, when you break this target down with the mindset to develop a clear pathway to what can be achieved now and possibly in the future, you will feel much more confident.

Like many other sustainability initiatives, such as 'zero waste' or water conservation, reaching 100% renewable energy involves developing a plan for the long term whilst at the same time making sure you improve today.

Your particular pathway depends on your preferences, the financial viability of the projects, organisational risk profile, technical feasibility, and capacity of the available energy efficiency and renewable energy opportunities. There is no one-size-fits-all, and the external environment continues to evolve, so you need to work out which pathway works best for you.

All throughout the chapters in Step 2 you worked hard to find suitable opportunities by analysing your energy situation, engaging your stakeholders, and investigating and prioritising your options. In this chapter, you will conclude your planning work by packaging the opportunities in a 100% renewables pathway and developing an action plan.

To give you an idea of how other organisations have approached their journey, included are several case studies and various ideas on how you can visually display your pathway.

Building your pathway: reduce – produce – purchase

Depending on their location, most organisations could be 100% renewable from the next energy bill, simply by deciding to buy all of their electricity from renewable sources and purchase carbon offsets for stationary fuel sources and transport emissions. However, buying renewable energy from your retailer is usually more expensive than conventional power and represents a straight cost to your business.

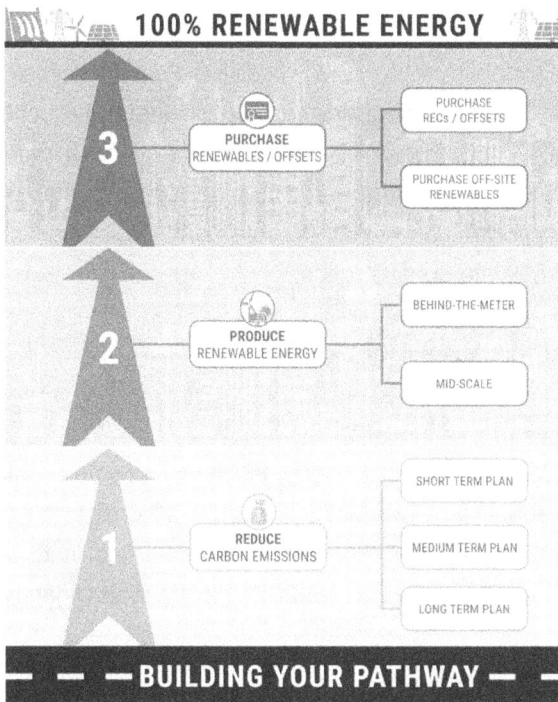

Figure 26: Building your pathway to 100% renewable energy

It does not deliver additional benefits like the satisfaction of producing electricity on your site or reducing the underlying energy demand. It also does not give you greater control over your energy use, allow you to innovate, or provide you a return on your investment.

Achieving 100% renewable energy is increasingly seen as a journey, with several step changes along the way. Based on my experience, the options to get you to your end goal can be organised into three opportunity areas: **reduce**, **produce**, and **purchase**.

Reduce your energy demand

A path to 100% renewable energy should always consider projects that focus on your organisation becoming more energy efficient. (See Chapters 3 and 8 for more on this.)

Being more energy efficient means you do not have to install as much renewable capacity, which is better from an environmental perspective. Energy efficiency is also, in some cases, better from a return-on-investment standpoint, which is why I call it the cheapest fuel.

In addition to having short-term goals for energy reduction in your pathway, like lighting upgrades, you should also have medium- and long-term strategies. For instance, your planned capital improvements, end-of-life upgrades, and fleet procurement decisions should take energy efficiency into consideration.

Produce your own renewable energy

Behind-the-meter

While you work to improve energy efficiency, you can ramp up your on-site renewable energy production. An organisation that has committed to 100% renewable energy should try to generate as much renewable energy on, or near, their site as is technically feasible and financially viable.

Energy produced onsite completely displaces energy from the grid and is produced locally at the point of consumption. This means there are no transmission and distribution losses that occur in the network once electricity is transported over a distance.

Having renewable energy installations onsite may come with the additional benefit of being visible to staff and visitors, which sends a clear signal to your internal and external stakeholders that you are concerned about the environment. Google, for instance, despite being one of the biggest purchasers of renewable energy worldwide, has installed a huge 1.9-MW solar array at its main campus in California that meets 30% of the site's peak energy use.[184]

On-site renewables generation is usually dominated by solar PV systems behind your meter that are mounted on your roof, carports, or on the ground. Small-scale wind power, geothermal, bioenergy, micro hydro, and solar thermal are options as well (see Chapter 9). In future years, you may also increase your renewable energy generation potential with battery storage.

As an average, it is likely that solar systems behind your meter can supply 15–25% of your total electricity consumption, subject to sizing limitations, available roof space, and the energy intensity of the facility.

Organisations that have already installed solar panels are usually surprised to find their installations only supply a low percentage of their overall electricity consumption. I have worked with municipalities that have solar PV installations on their administration buildings and libraries, but they only contribute 1–2% to the overall total energy consumption. A club I worked with has a 99-kW system on their roof that supplies a mere 3% of their total electricity consumption due to the energy intensity of their facility.

Very energy-efficient commercial buildings may be able to offset all their energy consumption with an equal amount of renewable power generation on-site. Warehouses, wholesalers, and large retailers may find they can meet most of their daytime demand with renewables if they have a roof with optimal conditions for solar.

Buildings with a high demand load, like hospitals, may not be able to generate all their electricity needs from behind-the-meter installations unless they have a lot of space and can mount renewable energy installations on the ground. Manufacturing organisations tend to sit in the same boat and won't be able to meet their entire consumption simply by putting solar panels on their roofs.

On your property, in-front-of-the-meter

This is where mid-sized renewable installations come into play. You can install a larger renewable energy system in front of your meter, with the generated energy offsetting your retail power purchases. Subject to available space, the system could be sized large enough to meet the aggregate consumption of all your sites.

Sizing the installation large enough also means you do not have to buy renewable energy elsewhere. To see whether this is attractive, compare it with other options, like entering a purchasing agreement, then evaluate the cost of energy generation, the treatment of the renewable energy certificates that result from the project, and negotiate this additional power supply source with your retailer.

Organisations that install intermittent renewable energy generation, like wind or solar, often worry how they can be 100% renewable if the sun does not shine or the wind does not blow. They also wonder how reaching their goal includes energy efficiency measures for appliances and equipment that may be powered by fossil fuel sources in the interim.

One example that seems to be a particular sticking point for local governments is the question of how to make infrastructure such as street lights more energy efficient in a way that contributes to their renewable goal. Street lights are run at night, they connect to the grid, and council is unable to control how they are powered.

The answer to that question is twofold. Firstly, it is hard to reach 100% renewable energy all at once – usually it is a long-term journey. Secondly, 100% renewable energy is achieved by balancing consumption with the equivalent amount of renewables via net use accounting, not by meeting real-time demand with real-time renewable energy production. So, before you make all of your power supply renewable, you must bring down demand as much as you cost-effectively can.

Purchasing renewable energy

Once you have exhausted all the cost-effective opportunities to reduce energy consumption and produce renewable energy on-site, you can consider *purchasing* renewable energy or renewable energy certificates (RECs) to make up the remainder of your energy consumption. Buying renewable energy may also be a good option if your organisation leases space rather than owns it.

The traditional way for energy users to source renewable energy has been through the purchase of green power from energy retailers. Usually, this attracts a premium to grid-supplied electricity. Similarly, organisations can buy RECs at the prevailing market rate.

Purchasing carbon offsets

Where things do not go as planned, where your energy projects fall short of your desired targets, or are delayed, carbon offsets may be a good way to ensure you meet your goals whilst maintaining the target pathway to 100% renewables. This may be especially useful where you have included transport or stationary fuel sources in your boundary and where you cannot readily transition to renewables.

Visually displaying your pathway

There are various options to display your pathway. You can present a table that lists the opportunities, their capacity to reduce grid energy, their costs and benefits, and your approaches to implementation. In addition, the table will contain the timing of the implementation. Some of your opportunities will be ready to go; some you may need to park for future development. This leads to your pathway being made up of short-, medium- and long-term opportunities.

You can also use an area graph that illustrates how fossil fuel energy consumption reaches zero in your target year with the implementation of your initiatives. Conversely, you could display a graph that shows how your energy efficiency and renewable energy measures grow till they meet your energy demand.

Some organisations like to use marginal abatement cost curves (MACCs) to display opportunities based on the financials of a business case and how much they can contribute to the goal. Just like MAC curves are used for greenhouse gas reductions, you can show the cost of your energy reduction, production, or purchase in cost/MWh at a particular point in time.

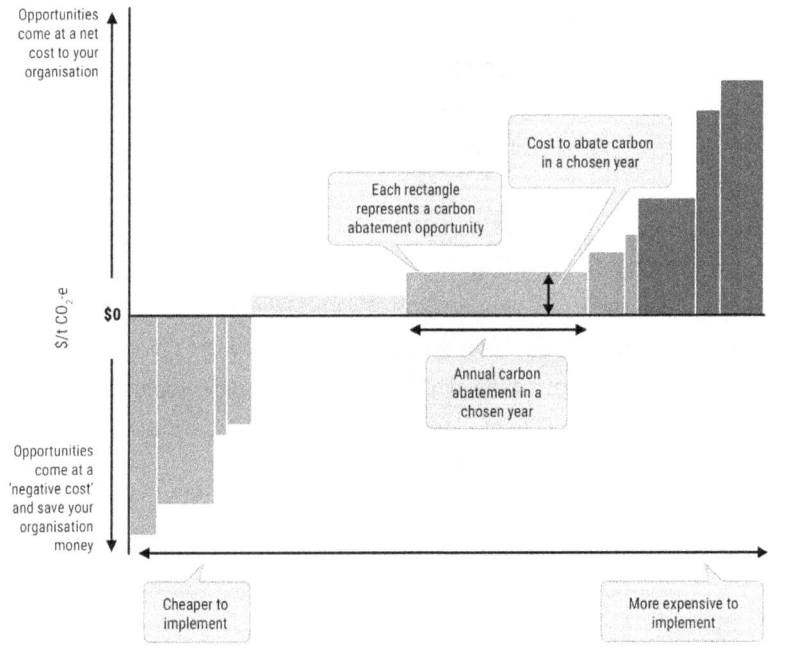

Figure 27: Example of a MACC, a marginal abatement cost curve

The vertical axis shows the money that needs to be spent per MWh abated, whereas the horizontal axis illustrates the size of the abatement in MWh. How far the various energy opportunities extend to the right shows how big the potential MWh reduction can be.

The graph is ordered left to right, from lowest to highest cost opportunities. Energy projects that return money to your pocket are depicted below the horizontal (zero cost) line. Projects above the horizontal line cost you money to implement.

If you seek inspiration on how to present your particular pathway, you can find several examples my team and I developed online at www. barbaraalbert.com.au.

A few pathway examples

BMW Group is a global German-based automobile, motorcycle, and engine manufacturer, which is aspiring to be the most sustainable company in its industry. By 2020, BMW Group will source 100% of its electricity from renewables. The group deploys a mixture of energy efficiency measures, on-site generation, and purchasing renewable power from its retailer. At its Leipzig production facility in Germany, for instance, four on-site wind turbines power the entire manufacturing process.[185]

BT Group is one of the world's largest communications services companies, with a strong focus on sustainability. The group has a goal to procure 100% renewable electricity globally by 2020. Its pathway consists of a mixture of off-site generation and purchasing green power. BT Group invested millions of pounds in three wind farm sites in the UK and purchases its energy via PPAs. In other countries where it is present, it buys green power from its retailer.[186]

IKEA Group is a home furnishing company with 336 stores in 28 countries. The company has committed to producing as much renewable energy as the total energy it consumes in their buildings by 2020. As of 2016, it has allocated US$1.9 billion in investments in new renewable energy projects. IKEA prefer to own both on-site and off-site renewable energy assets, and has installed more than 700,000 solar panels on IKEA buildings around the globe. It has purchased wind farms in Canada, Ireland, and the US, and own and operate a total of 279 wind turbines as of 2016.[187]

Lismore City Council,[188] in Australia, has committed to self-generate all its electricity from renewable sources by 2023. It plana to achieve this by continuing its energy efficiency efforts, installing solar PV behind the meter, and augmenting its installations with battery storage once it becomes cost effective. For the remainder of its electricity consumption, the strategy is to build a mid-scale solar PV project.

Apart from organisations that are still on their journey to meet a 100% renewable energy goal, let's look at two case studies of businesses that have already attained this status.

Microsoft has achieved carbon neutral and 100% renewable energy status since 2014 through a mixture of on-site and off-site installations, as well as through purchasing green power. It signed two PPAs for wind projects generating up to 110 MW annually, installed solar panels at its Silicon Valley campus, and purchase RECs for the remainder of its electricity demand.[189]

The sustainability journey of the Green Brewery[190] in Austria started with a big drive for cost savings. Rather than cut staff costs, the brewery undertook a meticulous examination of its energy profile to investigate where the most significant inefficiencies were.

This led to several energy conservation projects, like recovering close to all of its waste heat. It also constructed a plant that uses the brewery's wastewater to produce biogas, which partially replaced the brewery's natural gas consumption.

With one project after another, it substituted its natural gas consumption with alternatives until it was able to eliminate half of its natural gas consumption. Seeing that success was possible, it eyed a stretch goal, which resulted in committing to cover all of its energy demand from renewables and eliminate its natural gas consumption.

Not far from the brewery sits one of the biggest sawmills in Austria, which uses its wood processing waste in a biomass boiler for heat production for its drying processes. The brewery and the sawmill entered a strategic partnership to share heat output, and the brewery can now replace an even bigger portion of its natural gas consumption.

While it is important to develop a strategy to achieve 100% renewable energy, sometimes the unforeseeable – good as well as bad – happens. Organisations might approach you with local solutions you were unaware of to resolve your problems. New information might become available, or you might uncover opportunities you did not consider previously. You benefit when your plan is flexible enough to accommodate changing circumstances.

With the support of holding company Heineken, the Green Brewery installed a solar thermal plant. However, there was still residual natural gas consumption. With its ambitious goal of zero natural gas use, the brewery undertook further investigations to find potential solutions.

This resulted in the construction of a fermentation plant that uses spent grains as an input to produce biogas. Excess biogas is used in a cogeneration plant to generate renewable electricity, and the digestate is used as a natural fertiliser for agricultural businesses. Rather than own the biogas plant, the brewery entered into a 16-year, design-build-operate contract with a third party.

The brewery purchases accredited green power from its retailer for all its electricity needs. As a result of the brewery's relentless efforts to cut their ties to fossil fuels, it has now achieved 100% renewable energy status for both its electricity and stationary fuel consumption.

Translate your pathway into a plan

Preparation work

Once you create the high-level pathway for your 100% renewable energy goal, it is time to consider your existing infrastructure and which barriers need to be overcome to turn projects into reality.

Let's take the example of electric vehicles. If you plan to introduce EVs to your fleet, you need to think about the implications for your future infrastructure.

- If the vehicles will be charged on your site, can your renewable energy production cover the additional load?
- Will you have a separate charging station network?
- Will you allow vehicles outside your corporate fleet to charge their vehicles?

- Will you ask your employees to reimburse you to charge their private vehicles?

Some of your opportunities may also encounter regulatory issues and other barriers, so it is important to plan how to overcome them. Problems and barriers may relate to exporting energy, connecting to the grid, or a future potential for virtual net metering. Alternatively, they might relate to biofuel quality and availability in the medium to long term.

Once you have addressed these barriers and infrastructure considerations, you can start to develop your action plan, which will translate your pathway into a plan you can then implement.

Develop an action plan

One of the last planning instruments in Step 2 is the creation of an action plan with tangible goals, owned by individuals that will move your program towards fruition. An action plan is a written document that aligns with your pathway and your organisational directive.

Consider having your action plan as a separable document from your pathway to 100%. Why? Your strategy will likely not change as much as your action plan, which you will probably review once or twice a year.

Your action plan bridges Step 2 – 'Plan' and Step 3 – 'Implement'. You do not have to map out every single action required to get you to your end goal; specifying actions for the next couple of years will suffice.

Your action plan enables your pathway, and you may have several action plans that support your short, medium and long time frames. Your short-term action plan will be very specific and list the various projects you need to undertake on a site-by-site basis and per financial year.

Typical contents of your short-term action plan will be the time frame within which each project needs to be executed, the resources to allocate, people accountable for specified actions, due dates, and how the success of the projects will be measured and evaluated. The key is to stage the projects so that the most cost-effective ones are deployed first, whilst you wait for other projects that do not (yet) meet your investment criteria.

Your medium- and long-term action plans may not be as detailed as your short-term one, because they may contain opportunities that only become financially viable as the costs of certain renewable energy technologies reduce.

Bear in mind that your action plan will be updated often, as new information becomes available, so as an immediate step, have senior management sign off on your action plan(s) and have a number of projects ready to be developed into business cases. (You will finalise your business cases during Step 3.)

Your checklist:

You may be able to delegate or outsource these tasks.

☐ Structure your opportunities into ones that reduce energy consumption, can produce renewable energy, or where you buy renewable energy.

☐ Determine how much and when your energy efficiency measures can contribute to your target.

☐ Determine how much and when your onsite renewable energy generation can contribute to your target.

☐ Determine whether and when you can potentially develop a mid-scale renewable energy project.

☐ Determine the residual energy demand in your target year and the amount of renewable energy you must purchase to become 100% renewable, if any.

☐ Visually display your pathway in a table, marginal abatement cost curve, or graph.

☐ Get your pathway approved by senior management.

☐ Develop an action plan covering the next few years comprising: projects that get you closer to your target, timelines, and people assigned to specific tasks.

☐ If needed, develop action plans for the medium and long terms.

☐ Get your action plan(s) signed off by senior management.

STEP 3 – IMPLEMENT

In Step 2 – 'Plan', you worked on building the pathway and associated plans that will see you reach 100% renewable energy by your target year. These projects must be incorporated every year you do your business planning.

When the strategy to move to 100% fails, it is often because it is poorly implemented. Many times, the problem is not the strategy itself but how it is executed, and how well senior management sponsors it, both in words and in action.

For your strategy to be successful, you need to make sure your actions are translated into distinct projects that move you closer and closer to your goal, in the order of priority you set out in Step 2.

Before you start to implement your projects, identify potential risks and how you can mitigate them to an acceptable level. You need to financially appraise and finalise your business cases so you can put them forward for management approval.

Proper governance and reporting to management is crucial, and having the proper management processes in place, including senior management support, and potentially a steering committee taking ownership of implementation, is critical to your success.

Pay attention to the organisational changes that your program introduces as well. Staff and third parties, like suppliers and contractors, need to know you plan to transition to 100% renewable energy. Personnel responsible for implementing projects need to be trained in the proper operation and maintenance of the new equipment, as improper operation or maintenance can reduce the amount of energy savings and, hence, compromise achieving your energy goals. Organisational policies may also need to change.

At the end of Step 3 – 'Implement', you will go to market and closely monitor the progress and performance of each project to ensure you achieve your desired outcomes within specified time frames and within the budget allocated.

Chapter 12

Managing your implementation risks

You have done a lot of work up to this point. You have analysed your energy baseline, developed a stakeholder engagement plan, prioritised your actions, and packaged them in a pathway alongside your preferred financing options.

However, it is important to remain flexible – things do not always go as planned. Sometimes, there will be changes to circumstances, some of them in your favour and others not. Nearly every project you implement comes with risks, which need to be identified and managed.

Risk refers to any factor that may adversely affect the success of your energy project in terms of realising project outcomes, achieving time frames, or meeting budgets. These factors include risks in your business environment that may prevent your project outcomes from being fully achieved.

Risk management includes risk identification, analysis, evaluation, allocation of responsibility, and risk treatment. The purpose of risk management is to mitigate potential threats that could challenge achieving 100% renewable energy in a structured way. The more prepared your organisation is in the beginning for potential impacts on individual projects, the higher the likelihood you will succeed.

Small projects may only need a brief scan for risks and opportunities, and ongoing monitoring as to when and if changes to the risk profile occur. Large and more complex projects should have a formalised system to analyse, manage, and continually report risks.

The easiest way to identify and evaluate risks is to run a workshop with a broad subset of the stakeholders you identified during Step 2. In this meeting, you use the collective experience of people in the room to find and discuss the risks most relevant to you.

Evaluate the risks in terms of their potential impact and likelihood of occurrence by using your corporate risk matrix, and workshop mitigation strategies for the severest risks. Risks that are low may be accepted by management without any treatment.

The following list details the risks my clients encounter most often. The list is not exhaustive, and you may identify other risks or opportunities in your workshop.

Financial risks

Actual costs and benefits differ from projections

Actual project costs can come in higher or, conversely, lower than expected. This may be due to inaccurate business cases or savings valuation, unforeseen project complications, incomplete scoping of a project, delays, and/or exchange rate fluctuations. Shortfalls below expected savings can also relate to latent conditions, commissioning problems, design flaws, and technology underperformance, as well as other factors.

To fully mitigate these risks, perform a rigorous energy analysis based on high-quality data, take a conservative approach to estimation of savings, and be sure to manage and monitor projects properly.

Energy price risks and opportunity

Future changes in commodity pricing for fuel, gas, and electricity, as well as potential changes to key aspects such as electricity network pricing, affect your energy project's ability to deliver the anticipated return on investment/ROI. Such risk can work for or against your energy efficiency and renewable energy projects.

Perhaps you calculated your business cases by applying current energy rates projected with a certain escalation rate. If so, risk might lie in having used too high an escalation rate when prices fail to rise that much.

If that happens, the business case for your energy projects looks better than it actually is, as lower energy prices translate into longer payback periods. Likewise, if your energy prices rise more than predicted, your projects will return more value than forecast.

To get a feel for the different scenarios, carry out a sensitivity analysis in your business case. For example, you can change input parameters, like the escalation and your discount rate, and see what the results are.

Purchasing green power and RECs

The price for RECs, renewable energy certificates, is driven by supply and demand, which means that in the spot (short-term energy) market, prices fluctuate. If purchasing RECs or green power is part of your pathway and the price goes up unexpectedly, the risk could be that you can no longer afford them, threatening the achievement of your goal.

I once delivered a presentation at a sustainability leadership conference about these same four steps to achieve 100% renewable energy. At question time, a sustainability manager mentioned his organisation had been 100% renewable for many years, until two years prior when the price for RECs suddenly doubled from four cents to eight cents per kWh. That increased price meant the company could not purchase as many RECs as in previous years, making them no longer 100% renewable.

Conversely, if you develop a mid-scale renewable energy installation and you *sell* RECs, you might experience an unexpected windfall in proceeds if the value of RECs rises.

Strategic risks

Reputational damage

The 'Dieselgate' scandal with Volkswagen in 2015,[191] where certain emissions controls were only activated during laboratory emissions testing, as opposed to real-world driving, highlighted the damage that can come from a false representation of environmental facts.

Be careful how you express your targets and any claims of progression against them. A lot of the strategies in this book are designed to help you mitigate this risk. On the flip side, if you communicate your goal and achievements well, the market will typically reward your efforts with positive media coverage.

Lack of senior leadership and commitment

If your strategy and action plan lack senior leadership or commitment, the success of your program will be doubtful, to say the least. Ensure you have their buy-in early on in the process and maintain it throughout implementation. Clearly articulating the direct relationship between your renewable energy target and the organisation's broad strategic objectives should significantly mitigate this risk.

Policies favour low upfront vs. total ownership cost

Not many organisations factor in the total life cycle costs of a particular asset or project; more often they focus on up-front capital costs of acquisition and overlook longer-term costs or benefits. When you pitch your business cases, be sure to show the financial appraisal of the total life cycle of your projects, with all the associated benefits, such as lower maintenance costs, improved health conditions due to upgraded air conditioning or lighting solutions, lower pollution resulting from fuel and gas energy efficiency projects, and/or increased productivity.

Changing market conditions

The energy efficiency and renewable energy market, along with supporting services, is driven by rapid innovation and technology changes. When you do not monitor new developments, you risk falling behind. Remaining alert to innovation will help you retain your leadership position.

Technology is superseded and improved

Some organisations delay implementing technology, because they think an updated version might supersede it. This wait-and-see attitude results in delays and the risk that you may lose out on savings by installing the technology now, rather than a few years down the track.

Regulatory risks and opportunities

Regulatory and administrative costs

Do you need approvals to implement your energy project? Have you factored in potential regulatory and administrative costs, or time and resources spent to access incentive schemes?

Early on in the process, contact your local planning authority, where applicable, to determine what kind of planning and building permits you might need. Your consultants can help you with this and consider the outcome in the design and feasibility stage.

Policy uncertainty

Policy uncertainty around support for renewables and low carbon technologies is a risk not to be underestimated. In your business case analysis, you may factor in government support, but be mindful that regulations may change as policymakers adapt to evolving markets and changing technologies, and learn lessons from previous regulations.

Government support comes in many forms. Some examples are guaranteed feed-in tariffs (FiTs), upfront subsidies, reduced fees to access consulting advice, grants, infrastructure provision (e.g. free charging stations for EVs or bike-riding support), tax exemptions, and preferential traffic treatment.

Many incentives have been scaled back or removed. For instance, wind and solar energy have grown much faster than policymakers expected, which led to the removal of subsidies like high guaranteed FiTs in some countries and regions. Similarly, policymakers have grown cautious in their support for biofuels, due to their potential adverse impact on global food security and the limited impact of biofuels on emissions reductions.

Countries also learn from each other. Germany pioneered modern FiTs in 2000 and, by 2010, 50 other countries had copied this model. Since then, Germany decided to phase out or reduce FiTs in favour of reverse auctions, which uses a competitive bidding process to achieve rapid development at low subsidy cost. The same situation applies to Australia: we used to have high FiTs, while now, government agencies are more supportive of reverse auctions (some of which are combined with FiTs).

Regulatory risk also manifests in laws and regulations regarding the network. The risk could be manifold: the transmission or distribution provider might not allow you to export more than a certain amount of power into the grid at any one time. The FiT you currently get for exporting electricity into the grid might be cut. Naturally, the flip side of the coin is an increase in the export tariff or the introduction of innovative solutions, like virtual net metering.

Operational risks

Personnel allocation

In many organisations, energy savings are not seen as significant compared to other business priorities and therefore not enough human resources are allocated to oversee them. The risk is that teams like asset or infrastructure departments are given responsibility to implement energy projects, but no single person owns the overall plan, which can result in an ad hoc rather than a coordinated approach to implementation.

Future changes in the organisation, like acquiring a new building or changed business processes, should be analysed for how they fit into the overall plan, including an assessment of the potential for on-site renewable energy generation and what room exists to maximise energy efficiency.

A good strategy to mitigate those risks is to appoint a person who has overall responsibility to implement the plan and who regularly reviews the progress against it. Another method to mitigate this risk is to ensure the organisational energy goal translates into operational and work plans and annual performance reviews, so monitoring and reporting responsibilities flow down through the organisation.

Lack of staff understanding, engagement, and commitment

If your staff does not understand why your organisation is changing the energy model and are not brought along the journey, you risk disengaging them and they may resist your energy efficiency or renewable energy initiatives. Unfamiliarity with the technologies often amplifies negative perceptions and adds an additional barrier. It also poses the risk of reduced energy savings from implemented projects.

For example, staff might dislike having sensors on lights, or having their screen go black after a few minutes of inactivity. They may not trust that an electric vehicle will drive just like a petrol vehicle, or agree with having solar panels on the roof. However, if you engage them and they understand why these projects are important, you are more likely to gain their support.

Another example is personnel hastily replacing damaged or faulty equipment without considering energy efficiency impacts. When you build your 100% renewable energy plan with a certain energy baseline underpinning it, new equipment that consumes much more energy than previous equipment may force you to adjust your plan.

Organisations that excel at reaching their 100% goal do so because they have engaged their staff so they feel ownership in the project. They do the right thing because it is in their own best interest to do so.

Action plan accountability and implementation disconnect

When organisations develop strategies and action plans for 100% renewable energy, the ownership usually sits with the sustainability, energy, or environmental team. Then, once the plan is developed, individual projects are implemented through the annual budget cycle.

From what my clients tell me, the sustainability department may have little authority to tender for, and manage, contracts that impact other teams or business units. A potential solution to this problem is to form a steering committee with ownership of, and responsibility for, the plan execution.

This way sustainability and other teams have a say in the execution of the plan. Additionally, they will know how the various organisational activities impact the 100% renewable energy goal, through regular meetings.

Business interruption

Any interruption to your usual business processes may adversely affect your processes, or the quality of your product or service delivery. It is critical that business continues to operate, even though an energy efficiency improvement is being implemented or your supply source changes to renewable energy.

This is why it is so important to carefully assess how a change might interrupt your business operations by consulting with relevant managers to put in place mitigation measures that reduce this risk. You need to pay even more attention to mission-critical processes.

Safety risks

If your project involves potential safety issues, then you need to conduct a safety assessment and show how the identified risks can be appropriately managed. You likely have an established safety management process that needs to be followed, one which may need to be reviewed or updated if your organisation develops projects outside your core area of operations.

Contractual and supplier risks

Once you have chosen a technology option, the next step is to engage the right implementation partner and set up appropriate contractual arrangements, like maintenance agreements or performance guarantees. Otherwise, you risk battling a non-performing supplier whilst you try to engage a new one.

You also want to make sure your supplier delivers on their promises and that, for example, an agreed two months' installation time does not turn into six months instead.

Technology risk

The technology you pick must be reliable, or at least serviceable, in case something goes wrong. Ideally, the technology is mature and well known, and has functioned for others.

At a conference I presented at, a speaker talked about how they had installed a co-generation plant sourced from the US. (The cogeneration manufacturer had no local presence in Australia.) The plant caught fire twice and in each case was down for six weeks, because spare parts had to be shipped from overseas.

If you can demonstrate to management that you have considered key project risks and identified strategies to mitigate them, your energy projects are more likely to be approved. The next chapter details the steps needed to finalise your business cases so that you can start to implement them.

Your checklist:

You may be able to delegate or outsource these tasks.

☐ Identify the potential risks of every project on your action plan.
☐ Apply your organisational risk matrix to evaluate the level of each risk, or apply a generic matrix, if you do not have one.
☐ Develop a risk mitigation plan for significant risks.

Chapter 13

Finalising your business cases

To ensure your organisation's limited resources are directed to the projects most likely to deliver high-value benefits, opportunities should be converted to viable business cases that can be approved and delivered.

The important function of a business case is to take data you have obtained, turn it into reliable business information, and pitch it in a way that allows decision makers to assess the merits of these opportunities.

During Step 2, you developed preliminary business cases with indicative costs, but those estimates could change at the detailed design stage or by the time you go to market. To make sure your projects do not fail financially, it is critical to calculate expected costs and benefits as accurately as you can. The best way to calculate the return on your energy projects is to examine their total lifecycle costs and income streams.

Lifetime costs and income streams

Energy projects cost money to install but result in lower operating costs throughout their lifetime. Energy project inflows are primarily savings through reduced expenditure, unless you receive external funds, like FiTs (feed-in tariffs).

While the direct benefits of your projects are lower energy bills and reduced carbon emissions, there may also be a range of non-energy-related financial benefits. If you change your lighting to LED, for instance, you will not need to replace globes as often as with previous incandescent or fluorescent globes, reducing maintenance costs.

Running air conditioning less translates to lower filter replacement costs. Reducing your energy consumption could mean you no longer require additional equipment you previously thought you would need. Having more daylight and natural ventilation in buildings might translate into healthier staff and increased productivity. Having your own renewable energy generation on-site might help you keep operating in case of a blackout, if it is designed that way. Incorporating whole-of-business benefits like these makes the business case for implementation more compelling.

Cash outflows, on the other hand, include initial capital outlay, which could include pre-project expenses, engineering and drafting costs, procurement, installation, project management, commissioning, staff training, and unavoidable lost-production costs.

They also include operations and maintenance expenditure, and principal and interest payments if money is borrowed. You may also want to factor in potential regulatory and administrative costs, or time and resources spent to access incentive schemes.

If you plan an energy project with lifetime government financial support, calculate cash flow with and without government funds, as one inherent regulatory risk is a change to policy. Under ideal circumstances, your business case will stand on its own two feet, even without the additional funds.

Financially appraise business cases

The bigger your project, the more accurate your estimate of costs and savings needs to be to ensure that money is invested wisely. Your finance department may be able to help you to financially appraise your business cases, or you can outsource this to a consultant.

The most common methods used to calculate the return on energy projects are simple payback period, net present value (NPV), internal rate of return (IRR) and levelised cost of electricity (LCOE), the latter being solely used for renewable energy projects. The NPV and LCOE methods are considered the most sophisticated and robust, whereas the payback period is the simplest to calculate. The payback period is most common, but for larger projects, the more sophisticated evaluation methods might be required. Let's have a look at how these calculation methods work.

Simple payback period

The simple payback period is the time it takes for the upfront cost of an investment to be recovered from the savings generated by it. The payback period is calculated by dividing the upfront cost by the savings per year.

Generally speaking, the quicker the payback period, the more beneficial it is for your business. If you can recover upfront costs quickly, the money becomes available again and can be used for other investments. You should consider investing in a project if its payback period is less than your target or acceptable payback period.

The spectrum of what organisations deem to be an acceptable payback period is broad. I have dealt with clients where the payback period had to be less than two years, and I have worked for clients who accepted any payback period for a sustainability project, so long as it was less than the lifetime of the asset.

Take the Green Brewery in Austria, for example. As mentioned in Chapter 8, on its journey to 100% renewable energy it was approached by a nearby saw mill that wanted to sell the excess heat from its biomass facility. The brewery investigated the cost of the business case – the capital cost to lay a private pipeline and the ongoing operational cost to pay the saw mill for the heat.

This cost was offset by savings in natural gas use. The resulting payback period was between five and six years, which – under normal circumstances – would not have met the minimum payback standard.

However, because the project met strategic sustainability goals, the management team accepted a longer payback period.

The advantage of the payback period is how simple it is to calculate and understand. Most organisations I deal with prefer the payback period method exactly for this reason. However, this metric does not consider the value of cash flows that occur after the payback period; nor does it take into consideration the time value of money (i.e., the discount rate).

Solar panels, for instance, have a useful life of 20 to 25 years. You might pay them back after six years, but after those six years have passed, you still incur savings over the continued lifetime of the panels, which you will not capture if you just apply the payback period.

Net present value (NPV)

If I asked you whether you wanted to have $100 today or $100 in a week's time, which option would you choose? The smarter answer, from a financial perspective, is to take the $100 today, as money loses value over time.

The net present value (NPV) method takes the time value of money into account and is ideal to appraise long-term projects. The further in the future a potential benefit or cost is, the lower its value. This concept is made tangible by a process called 'discounting'. You work out all anticipated costs and benefits of the energy project over its expected life and convert the value of a future return into today's value.

The NPV is the present value of all cash flows generated by an energy project. All cash inflows and outflows are discounted to the present value, using a target rate of return (also called the 'discount rate').

Take the dollar results of your NPV calculations and imagine someone hands you a cheque for that amount. Executing the project is worth that much money in today's dollars.

The discount rate is the return you need from an investment, and is based on the interest rate for any debt your organisation has, the investment return that your business requires, and the risk level of the project. Most local governments, for instance, apply a discount rate of 6–7%.

The NPV returns different results depending on the discount rate applied. The lower the discount rate, the higher the NPV. The selection of the appropriate discount rate is important to ensure that future project returns are not being over- or underestimated in today's value.

If the NPV is zero, the project pays for itself within the asset's lifetime. If the NPV is positive, not only will the project pay for itself but it will also generate a positive cash flow. If you must select between projects, or between financial delivery options of projects, choose the project with the highest NPV.

If the project has a negative NPV, you might still decide to go ahead with the project due to the co-benefits (like gaining reputation in the market). Alternatively, a few projects can be bundled into one, making the combined NPV look more attractive.

Internal rate of return (IRR)

The internal rate of return (IRR) is the discount rate that makes a series of cash flows have an NPV of zero. IRRs can be used to prioritise competing projects; usually, the higher the IRR, the more attractive the energy project.

Invest in projects with an IRR above the required rate of return. If your energy project has a return that is greater than your finance rate, you improve your cash flow, with little risk.

> If your energy project has an IRR greater than your company's profit margin, that energy project is the place to invest your money. Many energy projects have IRRs greater than 25%.

The disadvantage of the IRR is that it does not determine the value an energy project adds to an organisation. Two projects can have the same IRR but be of completely different sizes. For instance, if one project has an NPV of $200,000 and an IRR of 10% and another project has an NPV of $40,000 and an IRR of 10%, the project with the higher NPV is far more worthwhile for your organisation because of the extended asset life. Also, if you look at two mutually exclusive projects, the NPV provides a much more appropriate measure than the IRR.

> The payback period, IRR, and NPV do not have to be used exclusively. You can show these metrics in tandem to demonstrate the project's financial benefits.

Levelised cost of electricity (LCOE)

When it comes to installing mid-scale (or large-scale) renewable energy projects that are on the side of the grid (in front of your meter), you may choose to base your financial appraisal on a concept called 'levelised cost of electricity' (LCOE), or the levelised cost of gas (LCOG).

The levelised cost of energy is a sophisticated economic assessment of the total cost to build, operate, and decommission a power-generating asset over its lifetime divided by the total energy output over that lifetime. Calculating the levelised costs allows you to compare the financial merits of different energy-generating technologies of unequal lifetimes and differing capacities.

You can calculate the total costs by accounting for all of a renewable energy system's expected lifetime costs, including construction, financing, fuel and maintenance, taxes, insurance, and incentives. Adjust all cost and benefit estimates for inflation and discount them to take the time-value of money into consideration.

Typically, the LCOE or LCOG of renewable energy projects depends on the following factors:

- Initial capital, including all pre-project, installation, and commissioning costs
- Fixed operations and maintenance costs (e.g. fixed labour costs, equipment, annual preventative maintenance, insurance)
- Variable operations and maintenance costs (variable labour costs, consumables, repairs, spare parts)
- Replacement part costs (e.g. inverters and battery storage)

- Fuel costs (there are none with renewable energy projects such as wind and solar)
- Decommissioning costs
- Government incentives (e.g. a grant, carbon price, RECs, FiTs)
- Capacity factor
- Output degradation over time
- Discount rate
- Amortisation period.

The LCOE is usually expressed in cents/kWh, or $/MWh, whilst the LCOG is generally expressed in $/GJ (gigajoule). A low LCOE or LCOG means energy is being produced at a low cost, with higher likely returns for you. If the cost of a renewable technology is as low as electricity sourced from the grid, it is said to have reached 'grid parity'.

A renewable energy plant is initially expensive to build but has low marginal costs, since in most cases the fuel is free (like with solar or wind projects) and the maintenance is less than that of fossil fuel power stations. Marginal costs are the money spent to generate an *additional* unit of energy, over and above the fixed costs associated with the initial investment and operation.

The primary reason to calculate the levelised cost of your renewable energy project is to give you the minimum price at which you need to sell your renewable electricity/gas to the grid to break even. If you want to use the renewable energy you generate, then the price will be the LCOE of generation (plus any developer margin) plus the network costs and retailer margin.

Your renewable energy project will be financially viable if you can deliver the electricity or gas at an equal or lower cost to energy purchased from your retailer. As of 2016, it is likely that the price per unit of renewable energy delivered is higher than what you buy from the grid, but with further falling costs, this will soon reverse.

Accounting for changing circumstances through a sensitivity analysis

The calculated benefits and costs of an energy project may vary, depending on differing assumptions about the input data and methodology applied in your financial analysis (costs of installation, cost of maintenance, lifetime of the system, value of government incentives, discount rate, etc.).

To gauge the range of potential project outcomes for differing inputs, I recommend my clients perform a sensitivity analysis. One approach is to perform worst-case, most likely, and best-case scenarios, and to assign probabilities to each. This will provide you with a deeper understanding of which parameters are of primary importance, how they affect the viability of your opportunities, and where break-even points and other milestones lie with respect to changes in particular inputs.

Refining your business case pitch

Have you ever had an appointment with a doctor in which they explained your situation using medical jargon intermixed with Latin? Try not to use jargon and complicated terms when you present your business cases to senior management. If you purely focus on technical details, you risk disengaging the audience, which may result in your business cases not being approved.

> Avoid jargon and technical language. Pitch your business case in the language of the decision makers.

When you undertook your stakeholder engagement during Step 2, you uncovered the different views of key internal stakeholders/decision makers regarding your 100% renewable energy program.

When you pitch your business cases, make sure you address your primary stakeholders' concerns or potential arguments, especially of those who need to sign off on your business cases. The environmental benefits might persuade some people, but in most cases, people want to see the financial benefits.

Any energy savings that you can achieve via an energy efficiency or renewable energy project represents an existing waste stream that, by doing nothing, continues to drain operating cash unnecessarily. It is essentially a penalty you pay every month for failing to implement the project. So, make it clear how much money is being lost by not doing the project and show all the related benefits you can identify.

Other decision makers may be more driven by increased building values through higher sustainability ratings, or by the improved productivity, health, and safety of employees. Others may be attracted to the innovation being introduced, leadership in the market, or the marketing and positive public relations that energy projects may represent. It is important to find the right angle for your business cases so you can get them over the line.

Finalising your business cases

When the time for budget submission arrives in your organisation, you want to have your business cases costed and ready to go. When you finalise your business cases, familiarise yourself with the approval process in your organisation. There may be a particular template which you need to complete. The size and complexity of your energy project will likely determine how much detail you need to provide.

Your business case should briefly describe the objectives of the project and the process you followed to develop the opportunity. It is best to also explain the technical changes required and any relevant planning issues.

Specify the location and boundaries of the project, and the type of fuel being saved. Quantify the costs and benefits of your opportunity, estimate the direct and indirect savings versus business as usual, and recommend the best funding method. Your business cases should include how many tonnes of carbon emissions would be saved and how this would affect your emissions profile. Make sure you can substantiate your business case with documented assumptions and working sheets.

Your business cases should also demonstrate the potential risks of your projects and how they would be appropriately managed.

Also note the stakeholders involved in each business case. Ideally, your stakeholders have agreed on the deliverables of the project, including how their success will be measured. There also needs to be a clear assignment of project roles and responsibilities.

The next chapter covers more about these organisational changes, like establishing a management structure and setting up clear roles and responsibilities.

Your checklist:

You may be able to delegate or outsource these tasks.

☐ Calculate the expected return on your energy projects as accurately as possible.

☐ Perform a sensitivity analysis to gauge the range of potential project outcomes for differing inputs.

☐ Familiarise yourself with the approval process in your organisation.

☐ Pitch your business case(s) in plain language to management.

Chapter 14

Managing organisational change

A goal of 100% renewable energy can potentially result in lots of changes within your organisation. If you mostly buy renewable power to meet this target, then only a few departments, such as procurement, sustainability, and marketing, need to adjust their practices. However, if you intend to embark on a path of energy efficiency and onsite renewable energy production, more areas in your organisation will be affected, and over a longer period, maybe even a decade.

During any process of organisational change, people can experience confusion, anger, and uncertainty as they move through the transition. There may be a natural resistance to change because operational practices will be different, or a lack of understanding exists as to the overall process. To minimise this, plan for the change so the progress against your goal can happen as smoothly as possible.

Preparing for organisational changes may also reveal corporate preferences regarding the operations and maintenance of the energy initiatives you are about to put in. Some organisations prefer to manage this in-house, whereas others prefer to outsource these activities to a third party.

Many people become involved in managing organisational changes when transitioning to 100% renewable energy. Not everything has to be done by you! And you can always outsource the bulk of these tasks to a skilled consultant.

Establishing a management structure

In many instances I have observed how organisations create a strategy, communicate it, and expect the rest to happen by magic. It is important that you establish a management structure for your program, which identifies the stakeholders and their roles and responsibilities, as well as their accountabilities.

Senior management support

The ultimate responsibility and accountability for the program should be clearly defined and accepted at senior management level. Support from senior management is critical and should be visible to all staff in your organisation.

Ideally, your program has an executive sponsor who understands and accepts that the success of the 100% renewable energy strategy is their responsibility and truly owns the project, from idea to implementation. They should be willing to use their power and influence to support and defend the program, even when faced with opposition from colleagues or other stakeholders.

Setting up a steering committee

Apart from having an executive sponsor, you may benefit from having a steering committee in place. The composition of your steering committee should be carefully considered, as it is responsible for delivering the program.

It should be made up of a multi-disciplinary team for maximum impact. The committee should be involved in all of your organisation's major planning exercises, like purchasing decisions, expansion or divestment plans, or the redevelopment of existing assets.

Ideally, you have executives as part of the team, as well as representatives from supply chain, operations, communications and marketing, facility management, contract and fleet management, IT, finance, and human resources. The committee will approve business cases, monitor the program's success, scrutinise the budget, and take ownership of and communicate project risks. Your steering committee should meet regularly to keep track of issues.

For big projects, consider appointing a qualified project manager to drive implementation and report to the project steering committee. The project manager must be adequately supported and delegated the appropriate level of authority. This role can be assumed by your energy or sustainability manager, or another internal or external resource.

Reporting on progress

The steering committee needs to be regularly updated about the status and achievements of your 100% renewable energy program. Every time you communicate the status, begin by stating the milestones you reached in the last reporting period and those you will aim for in the next.

Updates regarding the project budget should also be provided. Explain what the planned expenditure was versus the actual, as well as the reasons for any deficits or surpluses. Your report will include any areas of concern, specific problems, or actions that need to be taken by the steering committee.

Progress towards 100% renewable energy should also be an agenda item on group meetings, which helps raise awareness and root the program in reality. Employees get an opportunity to ask difficult questions and voice their opinions. If progress towards the goal becomes a regular part of meetings, it helps to establish and maintain the momentum.

Sustainability coordination role

To support the transition to 100% renewable energy, your organisation will benefit from setting up a sustainability coordination role. This does not have to be a full-time position and you can assign the tasks to an existing position. You can also outsource these tasks to a skilled consultant. Typical responsibilities include having to:

- Deal with the energy projects' administration requirements
- Develop and track energy monitoring to facilitate internal and external reporting of energy use and cost information (e.g. input to future procurement, report against targets, inputs to external reports, working with an external data collection agency)
- Report on the energy programs, internally and externally
- Identify and target potential incentives or grants
- Work with business units during budget preparations to support the identification and justification for energy efficiency and renewable energy expenditure
- Engage with a range of external parties (governments, industry networks, suppliers, community groups, energy market participants) to link to potential future opportunities (e.g. battery storage, virtual net metering, peer-to-peer trading, community energy developments, new energy retailing models, etc.)
- Develop and deliver internal awareness-raising and education on energy sustainability
- Coordinate communication of your organisation's successes in energy sustainability and carbon abatement
- Source RECs or carbon offsets that meet organisational criteria.

Assigning responsibilities to staff involved with implementation

The achievement of 100% renewable energy is unlikely to be realised without widespread recognition within the organisation of what is expected of staff to support achievement of the goals, what skills are required, and what support management will provide.

The activities each person will undertake should be outlined, and it needs to be determined who is responsible for what, with reporting lines clearly defined. Everyone involved in implementing the strategy to 100% renewable energy needs to understand the specific outcomes that are expected, deliverables for their work area, their responsibility, and key performance indicators by which they will be measured.

Changing policies

With a goal of 100% renewable energy, your organisation needs to be as energy efficient as possible and consider renewable energy when new developments are undertaken. This means your existing policies and procedures may have to change.

Policies mostly affected are procurement- and asset management-related, like maintenance policies for optimisation. Asset management plans communicate the current condition of all facilities and aim to ensure the level of service is provided at the lowest long-term cost. Ideally, they should also seek to provide services with the least amount of energy consumption and maximise onsite renewable energy, where possible.

When budgeting for assets, a total cost of ownership approach needs to be adopted by looking at the overall costs and benefits over an asset's whole lifecycle. When new assets are built, energy efficiency and renewable power generation should be taken into account. For instance, it would be ideal if buildings were built facing north (south, for my readers in the northern hemisphere), both from an energy efficiency as well as solar PV perspective.

In your procurement policy, when energy-consuming equipment is purchased, ensure that energy efficiency is a consideration. It is important to establish procurement criteria for assessing energy use, consumption, and efficiency over the planned operational life of equipment or services. Also, plan for optimal replacement when new equipment is bought – which may not be a like-for-like one. For instance, if your electric hot water system fails, do not simply replace it with another, more efficient electric hot water system. Instead, consider switching to solar hot water.

When purchasing or renting buildings, you should consider their energy efficiency, as older buildings might not be as efficient as newer ones. Your fleet policy may also need to change to incorporate the principles described in Chapter 5.

You can also leverage existing maintenance and supply contracts with external parties, particularly mechanical and building services, by modifying KPIs to incorporate energy and carbon savings or reduction targets. The best time for these modifications is when contracts expire or when you go to tender.

Bring staff and suppliers along on the journey

Progress towards your goal ultimately relies on your people. Engaging, motivating, and supporting them on the journey is essential. Some of my clients have taken their staff on site visits to other organisations which have nearly or already achieved 100% renewable energy. This has proved to be a highly effective strategy to increase confidence of personnel that the goal can be achieved.

Maintain stakeholder commitment

Maintaining stakeholder commitment requires ongoing effort throughout the life of your program. This includes regularly reviewing the stakeholder analysis to confirm the assessment is still appropriate and identifying whether any new stakeholders have emerged in order to refine the key messages.

Make it a point to continue to communicate to your stakeholders by promoting the status of your program and what you have achieved thus far, and reinforcing and reiterating the project's benefits at every opportunity. Remind your stakeholders of the mechanism for them to provide feedback, and don't forget to respond constructively to issues they raise.

Think big but start small

Your vision of 100% renewable energy is bold and ambitious, but if you want to radically shift your organisation to the end goal, you will probably experience considerable resistance and setbacks. People generally prefer a more gradual approach. If you start with small pilot projects that prove the new way works and results in fewer costs, less maintenance, and a better way of doing things, people will find it easier to believe in and embrace your goal.

Choose pilot projects that provide immediate and positive results, and that come with minimal implementation risks. Small pilot projects give people a chance to develop their skills, find out what works and what does not, identify problems and overcome them, and, above all, discover that the new approach of doing things will be achievable, and not compromise business continuity.

One of my clients told me about their experience replacing direct online (DOL) with variable speed drives (VSDs). The benefits of upgrading to VSDs are that they can better align the work of the motor with the task or service being performed, which reduces energy costs.

However, for a long time, DOL drives were the norm across the industry, and people were uncertain that VSD drives would deliver the promised benefits. So, when the opportunity to replace the drives was brought up by management, staff initially resisted.

The only way they could be convinced was to undergo a trial project in which one DOL drive was replaced with a VSD drive. Staff were asked to gauge the maintenance and operating requirements, as well as the benefits. Once they agreed the trial project was a success, they were willing to replace the remaining drives.

Engaging project opponents

Communicating with stakeholders who resist or oppose a project can be difficult and challenging. They may be unwilling to engage with the project and can effectively undermine progress by spreading misinformation that can influence the views of other stakeholders.

In my experience, there are four main reasons that people or teams oppose energy projects:

1. People are afraid your initiatives will add to their workload or make it more difficult for them to perform their jobs
2. People are reluctant to operate a new, unfamiliar piece of equipment
3. People believe the solution is too expensive and argue it will not deliver the promised benefits
4. People are concerned that control is being taken away from them, and decisions are being made without their having a say.

It is important to try to understand their perspective, or what they perceive the project's potential negative impacts to be. Once you do that, you can identify any benefits or opportunities the project may deliver for them, such as any direct personal benefits, their opportunity to have a say in shaping the end results, or how they are 'doing their bit'.

Promoting information that challenges, invalidates, or addresses any threats the project poses from their perspective can help overcome opposition. It also helps to use the recommendations of external consultants to legitimise a particular project's approach. In addition, people want assurance they will receive the right amount of training to get accustomed to any new pieces of equipment.

Sometimes, your most vocal opponents can become your most effective supporters if you successfully include them on the journey.

Raising awareness

Moving towards 100% renewable energy will engage your staff because people can see the connection between renewable energy and climate change and see a tangible demonstration of what is possible.

Equally, your employees need to support your energy reduction initiatives. Your organisation is trying to be as energy efficient as possible, so energy awareness must become as much a part of your organisation's culture as safety, quality, and customer care.

For instance, let staff know how their behaviours influence energy consumption, and how they can contribute to reduce it. Examples are to switch off lights and equipment when they are not needed, being mindful of the energy consumption of newly acquired equipment, or to let the steering committee know about any organisational changes that might impact the energy baseline.

It is not just easier but also beneficial to focus your initial efforts on staff who embrace change, are passionate about your goal, and want to help drive your initiative forward. Early adopters will attract the first followers, and before you know it, most of your employees may be convinced of the benefits.

In your messaging to staff, stress the positive aspects of your energy initiatives, such as possibly their increased comfort through better lighting, heating, and cooling solutions. You may also be able to persuade staff of energy efficiency benefits if you can help them save energy at home or while on the road.

Your performance management system can be adapted to explicitly measure and reward people for their contribution to the 100% goal. Consider integrating your energy performance as a standard item in operational reports. If individuals and teams have achieved successes, they should be celebrated. You can also consider running competitions.

Raising the awareness of your 100% renewable energy goal is important for any new employees that join your organisation too. Your company's energy and carbon goals should be made clear in job descriptions and at induction, including individual responsibilities that significantly impact power consumption.

Employees need to be made aware of their responsibilities, but they should also be informed of the progress against the target, as we become more motivated when we see our efforts yield results. To keep things simple, report on the status of your various projects by using a traffic light format. You can also display the current percentage of energy reduction and renewable energy contribution versus fossil fuel-based energy.

It is smart as well to raise awareness levels with the people and companies your organisation deals with on a regular basis, like contractors or suppliers. They, too, need to be informed of their responsibilities. Putting requirements on suppliers can be a powerful way to achieve your goals.

In some cases, organisations that have publicly committed to 100% renewable energy have succeeded in having their supply chain follow in their footsteps. One example is Lens Technology, a major Apple supplier, which has committed to running its Apple operations entirely on renewable electricity.[192]

Training staff

It is important to have the right people and skills in your organisation to execute every aspect of your strategy. When new energy systems or technologies are installed, you cannot expect staff to automatically know how to operate or maintain them so that the anticipated energy savings are achieved.

Identify what training is required to equip your staff with the knowledge necessary to plan and implement energy management measures related to their jobs. Staff who directly deal with energy equipment need to be trained to follow the instruction manuals and the maintenance schedule set out by the manufacturer and take into account any instructions from the system installer. This is important to ensure the new energy equipment performs in accordance with specifications and expectations, to further ensure the desired energy savings are obtained.

A training-needs analysis will show where the gaps in knowledge and skills exist. All you need to do is investigate what is needed to complete the tasks in your 100% renewable energy plan, determining existing skill levels and evaluating the gap. Asset managers or maintenance technicians, for instance, may need to know how to troubleshoot on-site renewable energy installations.

If there is solar PV, for example, they need to learn the basic safe operation and proper maintenance of your system. They will need to regularly check whether it is still functioning as normal and know how to replace blown fuses, or how to reset inverters, breakers, and switches, if a problem arises.

Staff can learn the required skills by attending internal workshops, internet-based courses, or externally led training.

Remember: you can outsource the operations and maintenance of your energy systems if you prefer to not manage them internally. The added benefit of this is that it reduces risk and allows your organisation to focus on its core business activities.

It is good to choose your favoured option early, as you can modify your tender specifications accordingly. The next chapter discusses what you need to be mindful of when you go to market for your energy projects.

Your checklist:

You may be able to delegate or outsource these tasks.

- ☐ Get an executive sponsor for your 100% renewable energy program.
- ☐ Set up a steering committee.
- ☐ Put in regular meeting and reporting processes.
- ☐ Task someone with a sustainability coordination role.
- ☐ For bigger projects, appoint a project manager (which could be the sustainability coordinator).
- ☐ Assign responsibilities to staff involved with implementation.
- ☐ Conduct a stock-take of your policies and analyse which ones need to change to achieve your goal.
- ☐ Maintain stakeholder commitment by regularly communicating with them.
- ☐ Start with easy projects to get some immediate and positive outcomes, which hopefully results in a positive feedback loop.
- ☐ Engage your project opponents.
- ☐ Raise awareness of your goal amongst staff.
- ☐ Train the staff that will directly operate energy equipment (unless you choose to outsource this responsibility to a third party).

Chapter 15

Implementing and running your projects

This chapter mainly relates to implementing onsite energy efficiency upgrades and renewables. The process described here is different, as opposed to purchasing renewable energy through your retailer or through a PPA, a power purchase agreement.

After your business cases are approved and you plan the organisational changes to occur, you can go to market to select suitable implementation partners, like suppliers or project managers, for your projects.

Once your projects are implemented, your contractors will hand over the documentation so you can operate and maintain the equipment. However, you should strongly consider outsourcing these aspects to an experienced and skilled third party. Let's look at some of the issues involved.

Tender preparation

Before you can go to tender for your energy efficiency or renewable energy projects, you may need to tender for detailed design services or energy audits. You also need to prepare the tender specifications and plan for the contract management. You will gather all relevant contract documentation and then go to market. Bear in mind that your tender preparations must be consistent with existing operational practices.

Considerations for contract management

When considering your contract management requirements, give some thought to your planned delivery and financing model. Would you rather have multiple contracts for various components like engineering, design, construction, commissioning, and financing? Or would you rather have a turnkey project, where one provider handles all project aspects?

Consider who will manage the contract internally, how the contractor's performance will be monitored and managed, and what the contractor's reporting requirements are. You may need to check whether this is a skill you already have in-house or if it needs to be brought in. You should also check what organisational approvals and procedures apply, and how milestone payments will be structured. Your previous work in identifying risks and how to manage them throughout the course of the project is also important preparation work.

Gathering your information

When you go to tender, give bidders the background information about your sites and projects. Information that helps your suppliers prepare a proposal include past energy audits, your energy baseline, and detailed information about your sites.

For your specific project needs, include a technical description of your facilities affected by the project and your preferences – if any – regarding the make and model of equipment to be installed, such as solar PV panels, chillers, boilers, or lights. You may provide general drawings to show the proposed new infrastructure, plant, and equipment.

Ahead of your tender process for equipment provision, you may also need to go to market for detailed electrical, hydraulic and/or mechanical drawings for your proposed energy efficiency and renewable energy projects.

You also need to specify what standards need to be met and what your warranty expectations are. Together, all these documents and specifications will form part of your tender for equipment provision.

Going to market

Requesting proposals

Following the design and specification, you should request detailed quotations or tenders for the installation of your energy efficiency or renewable energy project, and a contract with your terms and conditions. Your organisation's procurement guidelines will specify whether you go to open tender or select tender, or appoint contractors you have an existing relationship with. The guidelines may also inform you as to how many quotes you need to seek, possibly depending on the contract value.

Based on detailed tender specifications and design, you will then conduct a formal request for proposal and invite suitably qualified parties to submit bids. It pays to do some background work to find companies that have done this kind of work before. It is a good idea to ask other organisations that have implemented similar projects for their recommendations.

> The clearer you are about what you want to achieve, the better the proposals you are likely to receive.

To make it easier to compare proposals, ask for a clear itemisation of the solution's component and installation costs, as well as specific responses to items such as experience, expertise, capacity, financial performance, warranties and guarantees, safety, training, maintenance, and monitoring, for example.

Plan a day during your request-for-proposal period when tenderers can come to your site to conduct a physical inspection to make sure bids are based on contractors with a sound knowledge of your site and needs.

Choosing a supplier

Once you receive the proposals, carry out an evaluation against a set of agreed criteria, which may include some mandatory (pass or fail) ones, as well as some that are scored against a rating scale. As a first step, check whether the bids are professional, address your evaluation criteria, and are sufficiently detailed. The following is a list of criteria to help you select a suitable bidder:

- Has a high-quality solution been proposed that will deliver the expected benefits?
  - What is included and excluded from the offer?
  - Has the supplier performed a detailed analysis of your energy situation?
  - Has the supplier offered quality products that are supported locally?
  - What are the warranty terms and conditions?
  - Are performance guarantees included?
  - Has system monitoring been included to allow you to check ongoing performance?
- Does the tenderer possess the experience, expertise, and capacity to deliver energy projects of similar technology and scope?
  - What reputation do they have, and what similar projects have they delivered?
  - What is their proposed project team and project delivery structure?
  - What project management capabilities do they have, including work, health, safety, and risk management processes?
  - Will the supplier conduct the work themselves or subcontract parts out? If so, what contractual agreements protect you?
  - Do the bidder and all sub-contractors have the necessary accreditations, certifications, and licences?
- What is the financial strength of the supplier? This is particularly important where you engage companies to build, own, and operate the equipment, as the supplier will likely need to cover

all or a portion of the upfront capital costs. Therefore, you want to be sure of the strength of their financial position by having their annual reports or financial statements examined by accountants or others with financial expertise.

- What is the proposal price and will this yield the expected benefits in energy and non-energy savings? A value-for-money assessment may be carried out that takes into account this aspect as well as non-price criteria. Bear in mind that the cheapest bid may not always be the best.

As a second step, it is common to shortlist a small number of suppliers for a detailed analysis and tender clarification. For your shortlisted bidders, investigate the proposals in greater detail by contacting referees and potentially visiting reference sites. Confirm the supplier has the capacity to deliver the project on time, and analyse further whether the proposed staff have the skills and knowledge required for the job.

You may also want to undertake a thorough background check on short-listed bidders to determine whether there are any matters which may adversely reflect on you doing business with them or on their capacity to deliver the work. Also check that the necessary insurances are in place.

It is important at this stage to also carry out a thorough check of each bidder's project financial analysis, to ensure that proposal financials can be compared on a like-for-like basis, and to conduct your own sensitivity analysis on input parameters.

You may receive proposals with diverse input parameters and assumptions, like different price escalators that achieve a certain financial result. You may need to undertake further modelling in order to properly compare the underlying economics of each bid. You may also need to re-run your cost benefit analyses to ensure that the project benefits you estimated are still obtainable when engaging the respective supplier.

Based on your rating criteria, the initial evaluation and follow-up checks, and the financial assessment, you can decide on your preferred bidder. Once you select your implementation partner, you can develop the final contract language and service plan, if applicable. For larger projects, engaging a supplier or contractor may mark the start of a long-term relationship, so consider them carefully.

Your organisation will likely have standard contracts they use for similar types of procurement. I recommend that you provide any standard contract documentation that forms the basis for your procurement contracts to potential suppliers along with other tender documentation. This way the bidder can flag any issues they might have with the contract, giving you an idea about how much negotiation and time cost will be involved, should you engage the tenderer.

Implementation

Implementation of your projects should follow the processes already used in your organisation to manage any building or asset improvement works. I recommend consulting with the departments in your organisation that handle the appointment and installation processes for further details on the contract and project management.

Commissioning

While your energy project is being installed, make sure you monitor progress against the stated deliverables and milestone dates, and routinely check that the installers conduct themselves professionally, with attention to safety and quality. After the installation, check the system works as intended via a thorough commissioning process.

Handover

Make sure you understand what products, services, support, and maintenance your contractors and other third parties involved intend to provide. After project completion, your contractors should train you in the new system and explain everything that was installed to the relevant staff, including shutdown and start-up procedures and maintenance instructions and schedules.

Rather than take care of these aspects yourself, you can engage a supplier to build, own, and operate the plant.

You should receive hard and soft copy manuals that describe, at a minimum, the system components, layout, warranty information, operating and maintenance instructions, start-up and shutdown procedures, performance guarantee details (if any), data monitoring details, equipment data sheets, contractor contact details, and any compliance certificates. Also make sure that as-built drawings are handed over which include the detailed design of the solution and all electrical drawings.

Operations and maintenance

Operations and maintenance (O&M) are the day-to-day activities necessary for your energy equipment to perform its intended function. Having proper O&M procedures reduces repair costs and unscheduled shutdowns, extends your equipment life, satisfies your warranty conditions, realises the predicted benefits through a constantly optimised system, and ensures your energy systems are operated safely.

Once your project is handed over to you, the best way to get all the above benefits is to follow the instructions as set out by the manufacturers in the manuals and to heed advice given to you by your installers. I recommend you establish or add O&M plans for each piece of equipment implemented and make these documents easily accessible.

Consider outsourcing O&M activities to an experienced third party.

Even if you outsource your O&M activities, it is a good idea to familiarise yourself with what these activities entail so you can get a better idea of what is involved.

Maintenance

There are three types of maintenance activities – preventive, predictive, and corrective.

Preventive maintenance

Preventive maintenance is done to increase the life of your equipment and help it run more efficiently. It encompasses routine inspection and servicing at intervals determined by equipment type, environmental conditions, and warranty terms.

Examples of good operations and preventive maintenance regimes are to make sure that batteries are not overcharged and discharged, which results in reduced battery life, and to regularly clean your air conditioning filters to avoid the system having to work harder due to reduced airflow. Inspect equipment like light sensors or inverters to see if they still work.

Make sure variable speed drives are well ventilated, as humidity and overheating can lead to premature failure. Regularly inspect and clean solar panels if they are not self-cleaning, like where panel tilt is minimal, or in dry or dusty areas.

As part of your preventive maintenance regime, you also need to renew those pieces of equipment with a lower life span than the overall system. With air conditioning equipment, for instance, the filters must be replaced periodically, as specified in the manuals, whereas the overall lifetime of your HVAC system might extend to 20 years. Inverters need to be renewed about every 10 years, whereas solar panels have a lifetime of about 25 years.

Predictive maintenance

Predictive maintenance is used to forecast when maintenance should be performed, which results in increased operational life and a decrease in breakdowns. It is typically done by specialised technicians and often involves the use of sensory equipment. An example of a predictive maintenance technique is thermography, which is used for electrical infrared inspections to detect excessive heat, load imbalances, and corrosion.

Corrective maintenance

Corrective or reactive maintenance addresses equipment repair after a breakdown has occurred. There is always a risk of unforeseen breakdowns, but a more proactive preventive maintenance regime can help mitigate those.

An example of corrective maintenance is if an inverter stops operating unexpectedly. This can be due to prolonged direct sunlight exposure or the build-up of salt, or dust. Most likely, there will be a delay before you notice your renewable energy system has stopped producing as much power as expected and you can send someone to have a look at the equipment. So, in addition to the cost of an unscheduled callout and potentially replacement parts, you run the risk of losing energy savings for a time.

To address the problem of downtime, it is necessary to be more proactive with regular inspections, whereas you can address the time lag from when you detect the system fault to fixing it with implementing data and alert management. The next chapter talks more about monitoring and tracking your equipment.

Your checklist:

You may be able to delegate or outsource these tasks.

- ☐ If needed, create a detailed design for your energy projects. (This is usually outsourced.)
- ☐ Determine the person/department to manage supplier contracts.
- ☐ Gather relevant information.
- ☐ Put tender specifications together and go to market.
- ☐ Select a suitable supplier.

☐ Get your energy equipment installed.

☐ Ensure there is appropriate handover upon completion.

☐ Make sure you can properly operate and maintain this equipment. (This is typically outsourced.)

Chapter 16

Monitoring progress and reviewing your plan

Your projects can be considered successful if you achieve your target outcomes on time, to the agreed quality, within budget, with risks appropriately managed, and stakeholder requirements met. With energy efficiency or renewable energy projects, the primary outcomes relate to a reduction in power use or the production or purchase of renewables.

How do you know if you achieve your target outcomes? This chapter explains how to set up proper measurement tools and processes so you can demonstrate project success to your stakeholders and provide updates on how you track against your 100% renewable energy goal.

If something is not going according to plan, or if you experience tremendous success with your energy projects ahead of schedule, you may want to adjust the plan developed during Step 2. The strategy or pathway on how you get to 100% renewable energy will likely change every few years due to technology changes, price reductions, shifting corporate strategy, expansion of operations, and/or increased energy use. However, you may need to update the action plan *yearly* with new information that becomes available.

Monitoring and tracking your results

Your organisation reaches the target of 100% renewable energy by balancing your annual power consumption with an equal amount in renewables, applying 'net use accounting' principles. As I mentioned in Chapter 2, the boundary is important for energy accounting, and you may have drawn your boundary around energy sources like electricity, other stationary fuels (like gas), or transport energy.

As you implement your energy projects, you either reduce your consumption or increase the proportion of renewables, or both. At minimum, at the end of every year you need to determine the following information across all your sites:

- The amount of energy consumed
- The amount of renewable energy
  - Produced
  - Consumed on-site
  - Exported
  - Purchased.

Ideally, you can also calculate how much energy you save with certain projects. Only with proper data management in place that tracks consumption and generation will you know how much closer to your target your energy projects have brought you, or how this year's energy performance compares to your baseline.

A client once told me how one of their solar PV systems was tendered for without having specified monitoring. The assets department wanted to reduce project costs and did not understand the value of adding internet communication ability to the system.

The sustainability department was not involved in the procurement and therefore did not know of the omission until after the installation, when they wanted to find out how much renewable energy the system generated.

For a potential low-cost solution check out LoRaWAN,[193] which provides wide area networks for the Internet of Things.

Electricity bills only showed the power consumed and what the solar PV exported to the grid; the sustainability management software only showed the net energy consumption; and no one in the organisation was tasked with regularly reading the actual performance from the inverters. What this meant for the sustainability manager was that they could only report the *estimated* generation of the system, with no way to know whether it actually produced that much.

This is why I recommend that, where practical, you make sure your energy projects have the ability to monitor their performance so you can track the output in your data systems. Every time you complete a project or a trial, monitor and verify the energy savings or renewable power generation to determine the success of your opportunity.

> Keep accurate records of your energy project's performance to compare before and after results.

Measured results allow you to identify what aspects of your estimates were accurate, and if the information used in your analysis was appropriate and of sufficient quality. You can then incorporate the key learnings of what was successful and which elements require refinement in subsequent forecasts and analysis.

With proper monitoring you might also be able to identify further opportunities for the operation and maintenance of your equipment, access up-to-date systems data, identify faults, and even determine billing irregularities and errors. If you can provide evidence of the benefits that energy projects deliver, successful projects can also provide you with a valuable internal marketing tool.

It can be difficult to directly measure your energy savings for **energy efficiency projects**, as savings represent the absence of power consumption. Instead, you can determine the savings if you compare energy use before and after you implement your initiatives, accounting for major influences like weather or changed uses.

One option is to apply the estimated savings from your business cases, but actual performance may differ from the forecast performance.

A better way is to compare the post-project performance against your business-as-usual forecast, with potential adjustments for changes, like increases in output, area, or employees. There are internationally accepted measurement methods, like the *International Performance Measurement and Verification Protocol*.[194]

Regarding **renewable energy generation**, your systems must be able to record the amount of renewable energy you produce, how much you use on-site, and how much you export to the grid. Ideally, the systems connect to a network like the internet, so you can query and download performance data at the click of a button, or through an app.

> If your implementation contract states a performance guarantee for your renewable energy system, then your implementation partners will make sure proper measurement tools are in place. They should give you a consolidated annual system performance report, that includes all performance aspects of your energy system and compares these with the guaranteed yearly performance.

If you have **battery storage**, you can record the flows from your renewable energy system to the battery, the flow from the battery to the energy end use, and the flows between the battery and the grid.

With regards to your **transport energy**, you can record, monitor, and review your vehicle fleet's energy use by fuel and vehicle type, number of vehicles, percentage of biofuels you use, and the proportion of electric vehicles in your fleet and their charging consumption. Some of these monitoring systems may already exist, particularly for combustion engine vehicles, but some may need to be established where new technology is being used, like for electric vehicles.

If you **purchase renewable energy**, record the renewable energy certificates you sold, the renewable energy certificates and amount of green power or carbon offsets you bought, and how many renewable energy certificates or carbon offsets you retired.

As mentioned earlier, depending on the results of your projects' actual performance, you may need to adjust the plan and timing you originally set out in Step 2.

Capturing your learnings

After you have completed energy projects, consider documenting your projects in a brief report or case study for distribution to stakeholders so you can capture the learnings for future projects. You can document what worked well and what could have been improved, whether the right preparations were made, or if the process could have been more efficient or effective, whether the right stakeholders were involved, and if everyone understood their roles.

Your report should summarise whether the targeted energy savings were achieved and disclose the project's benefits, final costs, and final payback result. It should also explain any variation in the actual and forecasted costs and benefits, and the reasons for these variations, so you can learn what you can improve in future projects.

Reviewing and revising the plan

Organisations change – they grow, they divest, new solutions enter the marketplace, the regulatory framework changes. Some projects may be easy to implement; with others you may run into barriers you did not even know you would face when you started out, which may result in delays.

One thing is certain: there will always be change, and your plan has to be flexible enough to cater for those changes. So, at least once a year, check progress against your target, review and incorporate estimates of resourcing requirements (both internal and external), and capture any learnings from your energy projects. Based on these findings you should then make the necessary adjustments to your plan.

The following examples illustrate the types of changes you might experience. One of my clients operates an asphalt plant that consumes a lot of energy to heat bitumen, a road surface material. After analysing the power consumption, load profile, and tariff structure of the plant, we recommended a solar panel installation. A year after, when the sustainability manager enquired with the site manager about the impending energy project, she found out that the old bitumen heating plant had been replaced with a new, mobile plant.

The new system required more energy than the old one, and so the site manager had also installed a generator to help with the electrical load. The extra generator used up the funds originally allocated for the solar panels and the setup with the new plant and generator also changed the site's energy baseline. The result? They could not proceed with the solar PV project that financial year and it had to be postponed to the next year.

Despite your efforts to communicate organisational changes to your staff, another classic example is when personnel replace damaged or faulty equipment in a rush, without considering energy efficiency impacts. If new, big energy-consuming equipment is brought online that consumes much more energy than the previous technology or plant, a plan adjustment needs to be made.

Every energy efficiency and renewable energy project you implement, no matter how small or large, moves your organisation further away from the unsustainable consumption of fossil fuels and closer to a clean energy future. With every additional project, your organisation becomes more and more experienced in their execution until it becomes integral to the way you do business, and a part of your organisational culture.

It does not matter if you have 'only' saved 10% in reduced fossil fuel consumption or carbon emissions. What matters is that you continue to pursue projects that save you the next 10%. You do not have to get to the 100% renewable energy target in one go, but it is important that you progress on your path to that goal.

Then, finally, you can claim the status of being a fully energy renewable organisation. You will celebrate this achievement in Step 4 – Succeed.

Your checklist:

You may be able to delegate or outsource these tasks.

☐ Make sure your energy projects allow you to monitor their results.

☐ Create a process that allows you to track the results of your energy projects over time.

☐ Every year, collate information on how much energy you consume, how much renewable energy you produce and purchase, and how much closer to your goal you are.

☐ After every project, produce a short report to capture the learnings.

☐ Adjust the action plan based on new information that becomes available.

☐ Get sign-off for your adjusted plan.

☐ Review your pathway at least every two years and apply any necessary changes.

Step 4 – Succeed

Congratulations, firstly, on coming this far in the book, and, secondly – and more importantly – on completing your journey to 100% renewable energy. Think back on all the hard work you have done up to this point. You learned about the most common energy efficiency and renewable energy opportunities available, including ones less common. You analysed your baseline, engaged your stakeholders, packaged the best options in a pathway, and steadily implemented projects until you arrived at your goal.

This step is about reaping the rewards of all your hard work. Let's talk about how you can communicate and share your journey with other organisations, so they can learn from your experience and follow your footsteps.

Chapter 17

Communicating and sharing your journey

Achieving the milestone

You started out by determining the boundary of your goal and drew it around energy forms like grid electricity, natural gas, or your transport fuels. You also set a target date by which you wanted to reach 100% renewable energy.

Imagine, now, that you are five or 10 years in the future. Year after year, your energy consumption decreased as you implemented energy efficiency measures. Your onsite renewable energy production grew as you installed more generation capacity and battery storage, and as new innovative solutions were enabled. You may have also purchased renewable energy certificates or carbon offsets.

Year after year, you compared your annual performance to the baseline year, as well as the year before, to see whether you were on track with your projects. You guided your internal and external stakeholder through the organisational change process to enable a smooth transition.

Every year, you compared your energy consumption with your renewable energy production or purchases and checked how much closer you were to your goal. And then, in your target year, you finally offset all of your power consumption with an equal amount of renewable energy – you have reached the status of a 100% renewable energy organisation!

Maintain your status

To maintain your special status, keep calculating your power consumption and comparing it to your renewable energy generation or purchase. If they are in balance every year, you can continue to claim '100% renewable energy'.

It is important to note that whilst you continue to use 100% renewable energy, it is in the context of an ever-evolving environment. Invariably, there will be changes, both internally, in your organisation, and through external drivers.

New people and stakeholders may come on board, your management or structure might change, or you may acquire a new business or divest one. There will be technology advances as geothermal, marine, electric vehicles, and biofuels become more feasible and as new solutions enter the market. One thing is clear – you must always innovate to stay ahead.

Hedging against price risk

Because you are committed to 100% renewable energy over the long term and must balance your annual consumption with renewables every year, consider applying clever financial strategies. For example, you can hedge against the price risk of changing values of renewable energy certificates, or carbon offsets, and use these market fluctuations to your advantage.

Your projects may generate renewable energy certificates, or you may purchase renewable energy certificates or carbon offsets. In a year where renewable energy certificates or carbon offsets trade at a low price, you can buy more than you need and bank them for future use.

Conversely, if you run a mid-scale renewable energy installation and the REC price is high, you can sell more of your renewable energy certificates to the market if your projects produce more RECs than you need to cover your own needs. Alternatively, you can retire any RECs from previous years, which you banked for future use.

Other risk management strategies were discussed in Chapter 12.

Make public claims about your renewable energy status

Being on the path to, or achieving a certain percentage of, renewable energy is a fantastic accomplishment, and you have every right to – and should — share the news of your achievements in your marketing materials. However, please be careful how you express your claims so as not to risk reputational damage.

Making inaccurate claims can backfire. Not only will your brand lose trust in the market, but you also risk being prosecuted for inaccurate claims.

Many countries around the world take a hard line on environmental claims. In the US, for example, the Federal Trade Commission (FTC) has the power to prosecute false or misleading 'green' marketing claims.[195]

In Australia, the Competition and Consumer Act 2010[196] states that businesses must not mislead or deceive consumers, or make false or misleading representations. Compliance with this law is maintained by the Australian Competition and Consumer Commission (ACCC).

According to the ACCC,[197] claims that are vague or cannot be easily substantiated are more likely to be misleading or deceptive. This is why it is important to make your claims specific, and clearly explain the actions your organisation has taken, in simple language.

For example, if you state that 'We have reached 100% renewable energy', but this only relates to one of your many sites, you risk misleading the market. If you claim 'We are powered by 100% renewable energy', but in fact you sold all the renewable energy certificates from your renewables generation, it is the buyer of these certificates who is eligible to make a claim to the renewable energy, not you. Even if you say you 'host' renewable energy installations, but you have sold all the associated renewable energy certificates, this might represent a misleading claim.

I recommend being very explicit in what you publicly state. You can detail whether you generated your own energy on-site, or off, whether you retained the associated renewable energy certificates or sold them, or whether you purchased renewable energy.

If you purchased renewable energy, you can state *how* you bought it – through a power purchase agreement (PPA), from your energy retailer, or through buying unbundled certificates, for instance. You should also consider disclosing the source of the energy, e.g. solar, wind, hydro, or biomass.

If you purchase less than all of your electricity from renewable sources, you should state the amount or percentage you bought. You should also state the period of energy consumption your purchase covers.

If you buy carbon offsets, consider mentioning where the projects are located, how they reduce carbon emissions, and whether they comply with any standards like the Verified Carbon Standard (VCS)[198] or the Gold Standard (GS),[199] for instance.

You should also reveal the boundary of your claim. Does it cover only grid electricity or include stationary and/or transport fuels? Is only one of your facilities covered, all facilities in a particular country, or all facilities in all locations? Finally, state the length of your company's commitment to buy or produce renewable energy.

If you are worried about making an inaccurate claim, consider working with an independent, experienced consultant to verify your claim. Part of this verification work involves calculating your overall consumption, contribution of on- or off-site renewables, and purchases of renewable energy and/or carbon offsets.

Celebrate and share your success

Achieving 100% renewable energy is a fabulous achievement and should be celebrated, as should reaching interim milestones, like 25%, 50%, or 75% renewable energy. This can be done through a formal ceremony (potentially as part of another event), organisation-wide congratulatory emails or newsletters, or via announcements on your websites and social media.

You can also let people know about your achievement via local or national newspapers, industry magazines, or through TV. The marketing possibilities are huge.

Each staff member who participated in the implementation of the project should be personally recognised for their contribution. You can also consider developing a 'wall of accomplishments' and locate it in a highly visible place.

Review the results of the stakeholder analysis you developed during Step 2, update it with any changes that have happened in the meantime, and make sure that key stakeholders know about the success of the program. Include relevant statistics, such as energy, greenhouse gases, and cost savings, as well as other business benefits. Give credit to those who helped you implement the program.

Promote the story to all your employees, so they know what you have achieved – but never promote something you are not sure about.

Enrich the communication experience. You can communicate reaching your 100% goal through a variety of means: presentations, workshops, meetings, newsletters, intranet updates, videos, bulletin boards, staff meetings, information sessions, or through private social media groups.

Employees and visitors tend to get excited about the outcomes of renewable energy installations. Some organisations I work with feature monitors in their foyers that display the current generation of their solar panels in real time. It is relatively straightforward to share what your solar panels or other renewable systems currently produce, how much clean energy has been generated over their lifetime, and how many tonnes of greenhouse gas emissions have been saved. Not only does this engage employees but it also demonstrates to visitors that you take sustainability seriously.

People connect renewable energy installations with efforts to mitigate climate change, but it is a bit more difficult to get the message of energy savings across. First, you need to have proper monitoring and reporting processes in place to know how much energy you save. Second, you must communicate these savings in a way that makes them tangible. One way is to compare it to something that makes sense to people, like cars taken off the road, or the avoided energy consumption of a number of average households.

Some of my clients also use 'black balloons', where each balloon represents 50 grams of greenhouse gases. Others compare the savings to the number of solar PV panels that *do not* otherwise have to be installed. Investigate what would make the most sense to your stakeholders and think about a good way to communicate that story.

Celebrate your success with stakeholders outside your organisation, like customers, industry associations, or regulators. You can write submissions to influence government decisions. The Australian Capital Territory (ACT) will be fully renewable by 2020, and you can consider lobbying your government for higher renewable energy targets to replicate the huge success the ACT has experienced.

You can engage your industry association to let them know about your journey and influence other businesses. You can also join partnerships or programs like RE100, if you have not already done so.

Businesses you should specifically try to influence are those that make up your supply chain. Once you achieve 100% renewable energy status, look outside your organisational boundary to see if you can get your suppliers to follow your footsteps. This increases general awareness and creates a snowball effect.

You can publish blogs, podcasts, or videos about your success story. You might also consider regularly running 'open days' or information sessions where you show the public through your premises and point out where you have improved your energy efficiency and installed solar panels, if it is safe to do so.

You can inspire and teach others how you arrived at 100% renewable energy to encourage and assist other organisations to do the same. You can achieve this via annual or sustainability reports, traditional media, and your website. You can also present at conferences or be a guest speaker at events. To gain maximum exposure, share your story through social media, your corporate email signature, and through your shopfront, if possible.

I once walked past my local Apple store and saw the following text displayed on their shopfront window: 'This store is powered by 100% Renewable Energy'. Upon seeing that, my immediate reaction to the brand was very positive. Apple does a fabulous job marketing its sustainability achievements.

When Apple launched its latest iPhone and iPad in March 2016, it announced that 93% of its facilities worldwide ran on renewable energy, including 100% of those facilities in the US, China, and 21 other countries. Apple also detailed some of its ongoing initiatives, including a 40-MW solar farm built in China, and an innovative solar project in Singapore.[200]

Another example of an organisation that is effective in sharing what it achieves is Google,[201] which has received multiple awards for its efforts, like the EPA's 'Green Power Leadership Award, and a top position on Greenpeace's 'Cool IT Leaderboard'.[202]

So that others can learn from your experience, consider sharing not just the projects you were successful with but also the ones that were challenging. In their quest to become fully renewable, the Green Brewery in Austria undertook a biomass conversion project, which was subsequently unsuccessful and cancelled.

In the wise words of Andreas Werner, the brewery master, 'The road to being a sustainability leader will not always be marked with success. There will be setbacks along the way, but what counts is to chip away at the target consistently and not to give up until you have reached it.'

Apply for an award

Awards create momentum, positive reinforcement, and are something people can be proud of. Once you make your achievement known to the market, you may very well be contacted to submit an application for an award. Alternatively, you can also browse available awards in relevant categories, such as innovation, sustainability/environmental, or local business awards.

Awards significantly lift the profile of your organisation and you can use them in your marketing materials, email signatures, web pages, social media, and display them in the foyer.

The Green Brewery in Austria won multiple awards for its achievements in energy efficiency and renewable energy. The following picture shows Andreas Werner, brewery master, in front of its business sustainability awards. The company has become an icon in the industry, with companies from all over the world travelling to Leoben to see the projects in action.

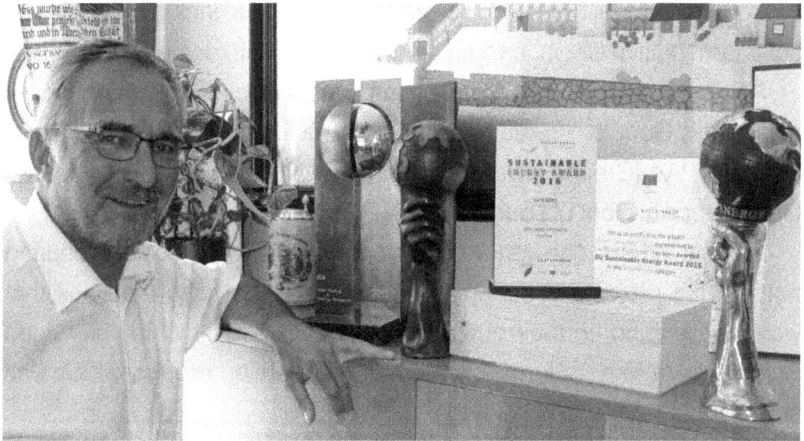

Figure 28: Andreas Werner, brewery master at the Green Brewery, Austria, in front of the business's multiple sustainability awards (photo usage courtesy the Green Brewery)

Lismore was the first regional Council in Australia to commit to 100% renewable energy and was also one of the first adopters of community energy projects. They won the prestigious Green Globe award in 2015 for their Renewable Energy Master Plan, as shown in the following picture. They were also selected as one of only two case studies to represent Australia at the Paris climate change talks in 2015.[203]

Figure 29: Left to right: Theresa Adams and Sharyn Hunnisett from Lismore City Council with New South Wales Minister for the Environment, Mark Speakman, at the 2015 Green Globe Awards (Photography by OEH/Lisa Madden)

Closing remarks

Reaching 100% renewable energy is both doable and achievable, but getting there may take you a few years. It involves an iterative, cross-functional journey, involving multiple departments, processes, and stakeholders.

For organisations willing to take on this challenge, the results are energy savings, locked-in long-term low costs, innovation, improved satisfaction of employees with the workplace, your organisation becoming a 'supplier of choice', attraction of ethical investors, and ultimately proof to the market that sustainability matters to you, which gives you a competitive advantage and raises your profile.

The Paris Agreement and the adoption of the UN Sustainable Development Goals made it clear that to keep global temperatures well below 2 degrees Celsius and to achieve zero net emissions, we need to deploy renewables rapidly. Decarbonising our energy supply will increase our global GDP and create millions of new jobs, along with new markets and new opportunities.

Switching your energy supply to 100% renewables places your organisation firmly at the forefront of the transition to a low-carbon economy and plays a vital part in creating a sustainable energy reality, one in which we can all prosper.

On a personal level, take credit for your efforts to help your organisation provide products and services that have been produced with clean, renewable energy, rather than fossil fuels that pollute our environment and contribute to climate change. You play an integral role in our quest to reduce carbon emissions and mitigate climate change through a better, more sustainable way of doing things. Your children and grandchildren will thank you for it.

Your checklist:

You may be able to delegate or outsource these tasks.

- ☐ From your target year, ensure that your annual energy consumption is met by an equal amount of renewable energy generation.
- ☐ If you sell or buy RECs, consider strategies to hedge against the financial risk.
- ☐ Make a public claim about the achievement of your status to promote it to the world.
- ☐ Make sure your public claim is transparent and does not mislead the market.
- ☐ Personally recognise each staff member involved in the program.
- ☐ Extend your reach to your suppliers and see if you can get them to commit to a 100% renewable energy goal.
- ☐ Lobby your government for higher renewable energy targets.
- ☐ Apply for awards.
- ☐ Give yourself a justifiable, big pat on the back.

If you enjoyed this book or found it useful, I'd be grateful if you would post a short review on Amazon. Your support really does make a difference, and I read all the feedback personally with the goal to make this book even better.

Industry abbreviations

AEMO	Australian Energy Market Operator
ARENA	Australian Renewable Energy Agency
AREMI	Australian Renewable Energy Mapping Infrastructure
BAU	Business as usual
BEV	Battery electric vehicle
BMS	Building management system
CCP	Cities for Climate Protection
CHP	Combined heat power
c/kWh	Cents per kilowatt hour
CFP	Carbon footprint
COP	Coefficient of performance
CPI	Consumer price index
CRI	Colour rendering index
CSP	Concentrating solar power
CSR	Corporate social responsibility
CST	Concentrating solar thermal
DNI	Direct normal insolation
DNSP	Distribution network service provider
DoD	Depth of discharge
DUOS	Distribution use of system
$/GJ	Dollars per gigajoule
$/MWh	Dollars per megawatt hour
EE	Energy efficiency
EPC	Energy performance contract
EPC	Engineering procurement construction

ESA	Energy services agreement
ESCo	Energy service company
EUA	Environmental upgrade agreement
EV	Electric vehicle
FCEV	Fuel cell electric vehicle
FiT	Feed-in tariff
GHG	Greenhouse gas
GJ	Gigajoule
GWh	Gigawatt hour (1 GWh = 1000 MWh)
HVAC	Heating, ventilation, and air conditioning
HW	Hot water
IEA	International Energy Agency
IoT	Internet of Things
IPCC	Intergovernmental Panel on Climate Change
IRENA	International Renewable Energy Agency
IRR	Internal rate of return
kVA	Kilo-volt-amperes
kW	Kilowatt
kWh	Kilowatt hour
kWp	Kilowatt peak
kW_{th}	Kilowatt thermal
LCOE	Levelised cost of electricity
LCOG	Levelised cost of gas
LED	Light-emitting diode
LGA	Local government area
LGC	Large-scale generation certificate
LPG	Liquefied petroleum gas
MJ	Megajoule
MRF	Material recovery facility
MSW	Municipal solid waste
MW	Megawatts
MWe	Megawatts electric
MWh	Megawatt hours (1MWh = 1000kWh)

NEM	National Electricity Market
NPV	Net present value
NUOS	Network use of system
O&M	Operations and maintenance
PHEV	Plug-in hybrid electric vehicle
PPA	Power purchase agreement
PV	Photovoltaic
R&D	Research and development
RE	Renewable energy
REC	Renewable energy certificate
REF	Revolving Energy Fund
ROI	Return on investment
SHW	Solar hot water
SLUOS	Street light use of system
STC	Small-scale technology certificate
STP	Sewerage treatment plant
Tonnes CO_2-e	Tonnes of carbon dioxide equivalent
ToU	Time of use
TWh	Terawatt-hour
VNM	Virtual net metering
VPPA	Virtual power purchase agreement
VSD	Variable speed drive

Online resources for further research

Bloomberg New Energy Finance

http://www.bloomberg.com/company/clean-energy-investment/

Bloomberg New Energy Finance is an authoritative source of data for clients, industry players, and the media on clean energy investment.

C40

http://www.c40.org/

C40 is a network of the world's megacities committed to addressing climate change. C40 supports cities to collaborate effectively, share knowledge, and drive meaningful, measurable, and sustainable action on climate change.

Ceres – Power Forward 2.0: How American Companies Are Setting Clean Energy Targets and Capturing Greater Business Value

https://www.ceres.org/resources/reports/power-forward-2.0-how-american-companies-are-setting-clean-energy-targets-and-capturing-greater-business-value/view

This report on Fortune 500 commitments is intended to inform companies, investors, the electric power sector, and state and federal policymakers on trends and preferences amongst large corporate renewable energy buyers. It is also intended to encourage companies in and out of the Fortune 500 to understand the value of setting renewable energy, energy efficiency, and greenhouse gas emissions reduction commitments.

Global 100% RE

http://go100re.net

Global 100% RE is a global initiative that advocates 100% renewable energy. Its goal is to initiate dialogue about 100% renewable energy, build capacity, and educate policymakers about the opportunities, case studies, and stories happening all over the world.

Go 100% Renewable Energy

http://go100percent.org/

The Renewables 100 Policy Institute created the Go 100% Renewable Energy project as part of its mission to study and accelerate the global transition to 100% renewable energy.

Greenpeace – *Clicking Clean* report

http://www.greenpeace.org/usa/wp-content/uploads/legacy/Global/ usa/planet3/PDFs/2015ClickingClean.pdf

Naming and shaming report by Greenpeace that evaluates the energy demand of the internet and the energy choices made by individual internet companies.

Greenpeace International *Energy [R]evolution*

http://www.greenpeace.org/international/Global/international/ publications/climate/2015/Energy-Revolution-2015-Full.pdf

This report by Greenpeace International shows a set of scenarios detailing how the world can get to 100% renewable energy by 2050.

RE100

http://there100.org/

Led by The Climate Group in partnership with CDP, as part of the We Mean Business coalition, RE100 encourage companies to set a public goal to procure 100% of its electricity from renewable sources of energy by a specified year.

Renewable Energy Buyer's Principles

http://buyersprinciples.org/

Facilitated by the WRI and the WWF, 51 corporate signatories developed these principles to spur progress on resolving the challenges they face when buying renewable energy, and to add their perspective to the future of the US energy and electricity system.

Renewable Energy Policy Network for the 21st century (REN21)

http://www.ren21.net/status-of-renewables/global-status-report/

REN21's goal is to facilitate knowledge exchange, policy development, and joint action towards a rapid global transition to renewable energy. The global status annual report provides statistics on worldwide deployment and investment of renewable energy.

Rocky Mountain Institute's Business Renewables Center

http://www.businessrenewables.org/

Founded by the not-for-profit Rocky Mountain Institute, the Business Renewables Center (BRC) is a member-based platform that streamlines and accelerates corporate purchases of off-site, large-scale wind and solar energy.

Sierra Club

http://www.sierraclub.org/ready-for-100

The Sierra Club is one of the US's largest and most influential grassroots environmental organisations. It is leading the charge to move away from the dirty fossil fuels that cause climate disruption and towards a clean energy economy.

Track 0

http://track0.org/

Track 0 is an independent not-for-profit that serves as a hub to support all those working to get greenhouse gas emissions on track to zero. It provides research, policy advice, communications, and networking support to governments, businesses, investors, communities, and NGOs.

We Mean Business Coalition

http://www.wemeanbusinesscoalition.org/

We Mean Business is a coalition of organisations working with thousands of the world's most influential businesses and investors. These businesses recognise that the transition to a low-carbon economy is the only way to secure sustainable economic growth and prosperity for all. To accelerate this transition, they have formed a common platform to amplify the business voice, catalyse bold climate action by all, and promote smart policy frameworks.

World Economic Forum, The Global Risks Report 2016

http://www3.weforum.org/docs/GRR/WEF_GRR16.pdf

The Global Risks Report 2016 draws attention to ways global risks could evolve and interact in the next decade.

World Future Council – *How to achieve 100% renewable energy. Policy Handbook*

http://www.worldfuturecouncil.org/wp-content/uploads/2016/01/ WFC_2014_Policy_Handbook_How_to_achieve_100_Renewable_ Energy.pdf

This 2014 policy handbook highlights eight case studies and offers policy recommendations for governments aiming to go 100% renewable.

WWF – *The Energy Report. 100% Renewable Energy by 2050*

http://wwf.panda.org/what_we_do/footprint/climate_carbon_energy/ energy_solutions22/renewable_energy/sustainable_energy_report/

Part 1 of this report spells out the main challenges to a 100% vision, and seeks to generate a discussion around the exhaustive scenario presented in Part 2.

Acknowledgements

There are so many people to thank for their help, support, kindness, friendship, and forgiveness along my path.

I am deeply grateful to my business partner, Patrick Denvir, without whom there would not be a Four-Step-Method to transition to 100% renewable energy.

I would like to especially thank all those who read this book in various drafts and gave me the benefit of their insights, experience, and constructive criticism, including: Michelle Dalmaris, Dr Peter Dalmaris, Jason Doyle, Sharyn Hunnisset, Professor Dr Stefan Jakubek, David Malicki, Nik Midlam, Joe Ritchie, James Robinson, Claudia Sattler, and Andreas Werner. Your input assisted so much to improve the quality of the final book and I want to thank you for the time you invested in this project, your honesty, and the support you provided me.

I also want to thank Alicia Bales, Neil Glozier, the Green Brewery, Herr Holzer and Herr Raser, Rob Kelly, Lismore City Council, the Office of Environment and Heritage, Sunshine Coast Council, Fiona Waterhouse, and Tim Wong, for sharing their case studies, photos, and general input into the book, as well as showing me around their renewable energy projects.

I am deeply grateful to all my valued clients whom I've had the privilege of working with over the years, which helped me perfect this four-step-method.

Glen Carlson, Andrew Griffiths, and David Dugan, you are each an inspiration, and have challenged and stretched my thinking and shown me that it is possible to realise my dreams.

David Longfield and Siobhan Gallagher of Longueville Media for your help with the book title, production, and editing, and Kym Latter and Werner Weißhappl, for your help with the book's cover.

I would like to thank my children, Marcel and Nico, for supporting me through this process, like writing me letters of encouragement and giving me lots of cuddles. Thank you to my parents and in-laws for all your support and love. Last, but not least, this book is dedicated to my husband, Christoph Strizik, who has guided and encouraged me, tolerated my endless hours of writing, and never stopped supporting me at critical times when it felt like I might never get to the finish line. I love you so much.

About the author

Barbara Albert is the founder of Sustainable Business Consulting and co-founder of 100% Renewables. She is a strategic adviser, author, and speaker, and works with businesses and governments on carbon reduction, climate change and energy management. She teaches several sustainability courses and is the go-to person for organisations interested in transitioning to 100% renewable energy or zero net emissions.

Barbara is a preferred supplier for the New South Wales and Australian Federal Governments and holds a master's degree in commerce from the University for Business Administration and Economics, in Vienna, Austria. She also studied at New York University's Stern School of Business, is certified in GRI sustainability reporting, and holds four sustainability and training-related tertiary certificates.

She is known for helping her clients achieve extraordinary sustainability outcomes that win them leading industry awards.

Barbara is married and lives in Sydney with her husband and two children.

If you would like more help and advice from Barbara Albert, please visit http://www.barbaraalbert.com.au.

There, you can order further copies of this book and download checklists, example pathways, and other great resources, like a quiz on your readiness for 100% renewable energy. There are also programs and online training available.

Notes and further reading

Websites referred to below were accessible at the time of publishing but URLs change over time so some may not be available in future.

1   https://www.weforum.org/reports/the-global-risks-report-2016/
2   http://www.iea.org/publications/freepublications/publication/WorldEnergy-OutlookSpecialReport2016EnergyandAirPollution.pdf
3   http://ethw.org/Milestones:Pearl_Street_Station,_1882
4   https://en.wikipedia.org/wiki/History_of_the_battery
5   https://en.wikiquote.org/wiki/Thomas_Edison
6   http://www.irena.org/remap/
7   http://www.ren21.net/status-of-renewables/global-status-report/
8   Ibid.
9   https://c311ba9548948e593297-96809452408ef41d0e4fdd00d5a5d157.ssl.cf2.rackcdn.com/2016-03-23-factcheck-renewables/ESA002-factsheet-renewables.pdf
10 http://fortune.com/2016/04/21/tesla-elon-musk-model-3-orders/
11 http://vancouver.ca/green-vancouver/green-transportation.aspx
12 http://w2.vatican.va/content/francesco/en/encyclicals/documents/papa-francesco_20150524_enciclica-laudato-si.html
13 http://www.un.org/sustainabledevelopment/sustainable-development-goals/
14 http://www.cop21paris.org/
15 http://unfccc.int/paris_agreement/items/9485.php
16 https://www.iea.org/newsroomandevents/pressreleases/2015/december/global-coal-demand-stalls-after-more-than-a-decade-of-relentless-growth.html
17 http://www.bloomberg.com/news/articles/2016-04-01/saudi-arabia-plans-2-trillion-megafund-to-dwarf-all-its-rivals
18 http://energy.gov/articles/6-charts-will-make-you-optimistic-about-america-s-clean-energy-future
19 http://www.irena.org/DocumentDownloads/Publications/IRENA_Measuring-the-Economics_2016.pdf
20 https://www.climaterealityproject.org/blog/follow-leader-how-11-countries-are-shifting-renewable-energy

21 http://denmark.dk/en/green-living/strategies-and-policies/independent-from-fossil-fuels-by-2050/
22 http://www.100-ee.de
23 http://wupperinst.org/en/a/wi/a/s/ad/2056/
24 https://www.sierraclub.org/sites/www.sierraclub.org/files/blog/RF100-Case-Studies-Cities-Report.pdf
25 http://there100.org/
26 https://www.cleanenergycouncil.org.au/policy-advocacy/reports/clean-energy-australia-report.html
27 http://www.enu.at/geschafft-100-prozent-erneuerbarer-strom-fuer-noe
28 http://www.ceres.org/resources/reports/power-forward-2.0-how-american-companies-are-setting-clean-energy-targets-and-capturing-greater-business-value/view
29 http://www.un.org/sustainabledevelopment/development-agenda/
30 http://www.un.org/sustainabledevelopment/energy/
31 http://www.un.org/sustainabledevelopment/climate-change-2/
32 https://www.unglobalcompact.org/what-is-gc/our-work/sustainable-development/global-goals-local-business
33 http://www.pwc.com/gx/en/services/sustainability/sustainable-development-goals/sdg-research-results.html
34 https://www.vestas.com/~/media/vestas/media/news%20and%20announcements/pdfs/globalconsumerwindstudy2012.pdf
35 http://cleantechnica.com/2016/08/18/apple-moves-one-step-closer-100-renewable-energy-worldwide/
36 http://divestinvest.org/2016-report/
37 http://www.greenpeace.org/usa/global-warming/click-clean/
38 http://media.virbcdn.com/files/f9/d6e716c56a9b3312-RE100AnnualReport2016_v17.pdf
39 http://www.thecro.com/wp-content/uploads/2015/10/Cost-of-a-Bad-Reputation-2015-Final.pdf
40 http://www.reuters.com/article/us-ukraine-cybersecurity-idUSKCN0VY30K
41 The other country not to sign the Protocol was the United States.
42 http://www.wemeanbusinesscoalition.org/sites/default/files/The-Paris-Agreement_Z-Card.pdf
43 https://www.virgin.com/richard-branson/the-paris-effect
44 https://www.youtube.com/watch?v=3rUPQ6u-nbc
45 http://www.industry.gov.au/Office-of-the-Chief-Economist/Publications/Pages/Australian-energy-statistics.aspx
46 http://www.minerals.org.au/resources/coal/national_dividends_of_a_strong_coal_industry
47 There are emissions associated with bioenergy use, but it is assumed that these emissions are re-absorbed by growing plants.
48 http://www.novonordisk.com/content/Denmark/HQ/www-novonordisk-com/en_gb/home/media/news-details.1968846.html
49 Information about eligible sources of renewable power generation as defined by Australia's Clean Energy Regulator can be found at http://www.cleanenergyregulator.gov.au/RET/Scheme-participants-and-industry/

Power-stations/Eligibility-criteria/Energy-sources-used-by-accredited-power-stations.

50 http://www.lismore.nsw.gov.au/page.asp?c=601
51 http://www.salesforce.com/company/sustainability/operations.jsp
52 http://www.pg.com/en_KE/sustainability/environmental_sustainability.shtml
53 http://www.industry.gov.au/Office-of-the-Chief-Economist/Publications/Pages/Australian-energy-statistics.aspx
54 https://www.sandiego.gov/sites/default/files/legacy/planning/genplan/cap/pdf/CAP%20Adoption%20Draft%202015.pdf
55 http://www.coffsharbour.nsw.gov.au/coffs-and-council/media-centre/Pages/2016/MR-Council-Adopts-Ambitious-Energy-and-Emissions-Targets.aspx
56 https://www.unilever.com/sustainable-living/what-matters-to-you/moving-to-renewables.html
57 http://www.cleanenergyregulator.gov.au/RET/About-the-Renewable-Energy-Target/How-the-scheme-works
58 http://www.ikea.com/ms/en_AU/about-the-ikea-group/people-and-planet/energy-and-resources/
59 http://www.industry.gov.au/Office-of-the-Chief-Economist/Publications/Pages/Australian-energy-statistics.aspx
60 https://www.environment.gov.au/system/files/resources/d67797b7-d563-427f-84eb-c3bb69e34073/files/100-percent-renewables-study-modelling-outcomes-report.pdf
61 The AEMO report envisaged that the same locations could be used for either photovoltaic (PV) solar or solar thermal (CST). If there was a combination of the two technologies, the amount that could be installed would be somewhere between these two figures.
62 http://arena.gov.au/files/2013/08/Chapter-10-Solar-Energy.pdf
63 https://www.cleanenergycouncil.org.au/policy-advocacy/reports/clean-energy-australia-report.html
64 http://newsroom.unsw.edu.au/news/science-tech/fact-check-australia-world-leader-household-solar-power
65 http://www.bloomberg.com/news/articles/2016-04-06/wind-and-solar-are-crushing-fossil-fuels
66 https://www.bloomberg.com/company/new-energy-outlook
67 http://www.bloomberg.com/news/articles/2016-04-06/wind-and-solar-are-crushing-fossil-fuels
68 http://www.cleanenergyregulator.gov.au/RET/Forms-and-resources/Postcode-data-for-small-scale-installations#Smallscale-installations-by-installation-year
69 http://www.cleanenergyregulator.gov.au/RET/About-the-Renewable-Energy-Target/How-the-scheme-works
70 http://reneweconomy.com.au/2016/construction-begins-on-australias-biggest-council-owned-solar-farm-74202
71 https://www.sunshinecoast.qld.gov.au/Council/Planning-and-Projects/Major-Regional-Projects/Sunshine-Coast-Solar-Farm
72 http://arena.gov.au/project/weipa-solar-farm/

73   https://www.cleanenergycouncil.org.au/policy-advocacy/reports/clean-energy-australia-report.html
74   https://www.iea.org/topics/renewables/subtopics/bioenergy/
75   https://www.cleanenergycouncil.org.au/technologies/bioenergy.html
76   http://www.cleanenergyregulator.gov.au/RET/Scheme-participants-and-industry/Power-stations/Eligibility-criteria/Energy-sources-used-by-accredited-power-stations
77   http://www.efa.com.au/Library/David/Published%20Reports/2005/Waste-ToEnergyGuide.pdf
78   Ibid.
79   http://www.cleanenergycouncil.org.au/technologies/wind-energy.html
80   https://www.cleanenergycouncil.org.au/policy-advocacy/reports/clean-energy-australia-report.html
81   http://www.bloomberg.com/news/articles/2016-04-06/wind-and-solar-are-crushing-fossil-fuels
82   https://ama.com.au/position-statement/wind-farms-and-health-2014
83   http://arena.gov.au/files/2013/08/Chapter-9-Wind-Energy.pdf
84   https://people.stanford.edu/cmakridi/sites/default/files/Makridis,%20Christos%20-%20Offshore%20wind%20global%20profitability.pdf
85   http://www.iea.org/publications/freepublications/publication/hydropower_essentials.pdf
86   https://www.cleanenergycouncil.org.au/policy-advocacy/reports/clean-energy-australia-report.html
87   http://reneweconomy.com.au/2016/tasmania-grid-struggles-with-drought-bushfires-lost-connection-95757
88   http://www.theguardian.com/environment/2008/jun/09/alternativeenergy.energy
89   Dispatchable energy can match supply with demand in real time.
90   http://www.powermag.com/australia-gets-hydropower-from-wastewater/
91   http://carnegiewave.com/
92   http://www.cleanenergycouncil.org.au/technologies/marine-energy.html
93   http://www.cleanenergycouncil.org.au/technologies/marine-energy.html
94   https://en.wikipedia.org/wiki/Geothermal_power_in_New_Zealand
95   https://www.ergon.com.au/__data/assets/pdf_file/0008/4967/EGE0507-birdsville-geothermal-brochure.pdf
96   https://en.wikipedia.org/wiki/Organic_Rankine_cycle
97   http://www.cleanenergyregulator.gov.au/RET/Scheme-participants-and-industry/Power-stations/Large-scale-generation-certificates/Buying-and-selling-large-scale-generation-certificates
98   http://go.sap.com/integrated-reports/2015/en/performance/environmental/energy-and-emissions.html
99   https://www.steelcase.com/content/uploads/2015/11/2015-Steelcase-CSR1.pdf
100   A quality criterion for GHG emissions reduction projects stipulating that the project would not have been implemented in a baseline or 'business-as-usual' scenario.

101 http://www.ceres.org/press/press-clips/can-you-power-a-business-on-100-renewable-energy-ikea-wants-to-try
102 http://www.wwf.org.au/what-we-do/climate/renewable-energy-buyers-forum
103 http://www.greenpeace.org/international/en/publications/Campaign-reports/Climate-Reports/Energy-Revolution-2015/
104 http://arena.gov.au/files/2015/11/ITP_REOptionsForIndustrialGas_Summary_Med_FA.pdf
105 http://www.cleanenergyregulator.gov.au/RET/Forms-and-resources/Post-code-data-for-small-scale-installations#Smallscale-installations-by-installation-year
106 http://arena.gov.au/files/2015/11/The-Green-Brewery.pptx
107 http://www.brauunion.at/ueber_uns/presse/presseaussendungen/offizielle-eroeffnung-solarthermie-anlage-unterstuetzt-brauerei-goess-auf-d
108 http://arena.gov.au/files/2015/11/ITP_REOptionsForIndustrialGas_Summary_Med_FA.pdf
109 Ibid.
110 https://www.publicworks.nsw.gov.au/sites/default/files/pdf/ESD_Publication.pdf
111 http://www.districtenergy.org/assets/pdfs/Webinars/Christchurch-5-11-15/2-ANDREW-BOUTCHARD2.pdf
112 http://www.ikea.com/ms/en_AU/pdf/reports-downloads/sustainability-strategy-people-and-planet-positive.pdf
113 http://arena.gov.au/files/2015/11/ITP_REOptionsForIndustrialGas_Summary_Med_FA.pdf
114 Ibid.
115 http://www.brauunion.at/ueber_uns/presse/presseaussendungen/offizielle-eroeffnung-solarthermie-anlage-unterstuetzt-brauerei-goess-auf-d
116 www.v-c-s.org/
117 http://www.goldstandard.org/
118 https://www.iea.org/publications/freepublications/publication/CO2EmissionsFromFuelCombustionHighlights2015.pdf
119 http://www.iea.org/publications/freepublications/publication/TrackingCleanEnergyProgress2016.pdf
120 http://www.mass.gov/eea/docs/dep/air/aq/health-and-env-effects-air-pollutions.pdf
121 http://www.irena.org/DocumentDownloads/Publications/IRENA_REmap_2016_edition_report.pdf
122 https://www.iea.org/publications/freepublications/publication/MediumTermEnergyefficiencyMarketReport2015.pdf
123 http://www.ikea.com/ms/en_AU/pdf/sustainability_report/group_approach_sustainability_fy11.pdf
124 http://about.hm.com/en/About/facts-about-hm/idea-to-store/logistics-and-distribution.html
125 http://www.theicct.org/info-tools/global-passenger-vehicle-standards
126 www.greenvehicleguide.gov.au

127 http://www.epa.vic.gov.au/your-environment/air/vehicle-emissions-and-air-quality/how-you-can-save-on-fuel-costs
128 Ibid.
129 http://www.cciqecobiz.com.au/assets/PDFs/CCIQ-ecoBiz-Fact-Sheet-Efficient-Driving.pdf
130 http://www.volvotrucks.us/powertrain/i-shift-transmission/i-see/
131 https://www.iea.org/publications/freepublications/publication/Global_EV_Outlook_2016.pdf
132 Ibid.
133 Ibid.
134 http://www.iea.org/publications/freepublications/publication/Tracking-CleanEnergyProgress2016.pdf
135 https://www.nzpost.co.nz/about-us/sustainability/electric-vehicles-powering-deliveries
136 http://about.bnef.com/landing-pages/germany-subsidises-evs-boosts-domestic-manufacturers/
137 http://www.transport.govt.nz/ourwork/climatechange/electric-vehicles
138 http://www.zmescience.com/ecology/renewable-energy-ecology/china-electric-bus-19012016/
139 https://www.daimler.com/products/trucks/mercedes-benz/urban-etruck.html
140 http://www.irena.org/DocumentDownloads/Publications/IRENA_Boosting_Biofuels_2016.pdf
141 http://www.irena.org/DocumentDownloads/Publications/IRENA_REmap_2016_edition_report.pdf
142 Ibid.
143 http://www.mynrma.com.au/motoring-services/petrol-watch/fuel-types.htm
144 http://www.afdc.energy.gov/fuels/biodiesel_blends.html
145 http://www.autoblog.com/2016/06/13/mercedes-benz-glc-plug-in-hydrogen-fuel-cell-coming-in-2017/
146 http://www.bmw.com/com/en/insights/technology/efficient_dynamics/phase_2/clean_energy/bmw_hydrogen_7.html
147 Quote by Michio Kaku, Author of 'Physics of the Future: How Science Will Shape Human Destiny and Our Daily Lives by the Year 2100'.
148 http://www.repositpower.com/
149 https://en.wikipedia.org/wiki/Pumped-storage_hydroelectricity
150 https://www.cleanenergycouncil.org.au/technologies/energy-storage.html#sthash.GLekLUk4.dpuf
151 https://c311ba9548948e593297-96809452408ef41d0e4fdd00d5a5d157.ssl.cf2.rackcdn.com/2016-03-23-factcheck-renewables/ESA002-factsheet-renewables.pdf
152 Noting that if there is a shift towards increased capacity and charges, the business case will be worse.
153 https://www.cleanenergycouncil.org.au/technologies/off-grid-renewables.html
154 http://www.aemc.gov.au/Rule-Changes/Distribution-Network-Pricing-Arrangements/Final/AEMC-Documents/Information-sheet.aspx

155 If you need to use diesel, you can balance your consumption by purchasing carbon offsets.

156 https://en.wikipedia.org/wiki/Vehicle-to-grid

157 http://cpagency.org.au/wp-content/uploads/2014/06/CPAgency_HowtoGuide2014-web.pdf

158 http://www.byron.nsw.gov.au/newsletters/2015/06/05/virtual-net-metering-media-launch-on-june-15th

159 https://www.smartgrid.gov/files/sg_introduction.pdf

160 http://www.energy.gov/articles/how-microgrids-work

161 https://www.energymadeeasy.gov.au/benchmark

162 https://www.environment.gov.au/climate-change/carbon-neutral/ncos

163 http://www.ghgprotocol.org/standards/corporate-standard

164 http://www.iso.org/iso/catalogue_detail?csnumber=38381

165 http://www.iso.org/iso/home/store/catalogue_tc/catalogue_detail.htm?csnumber=60088

166 http://infostore.saiglobal.com/store/Details.aspx?ProductID=1759652

167 http://www.oxfordcounty.ca/Portals/15/Documents/SpeakUpOxford/2016/100RE/OCDraft100REPlan20160622.pdf

168 http://www.oxfordcounty.ca/Your-Government/Speak-up-oxford/have-your-say

169 http://www.ikea.com/ms/en_AU/pdf/reports-downloads/sustainability-strategy-people-and-planet-positive.pdf

170 http://www.windpowermonthly.com/article/1380320/ikea-buys-424mw-project-finland

171 http://www.environment.nsw.gov.au/resources/business/financing-guide.pdf

172 http://www.environment.nsw.gov.au/households/solar-finance-guide.htm

173 https://www.unglobalcompact.org/

174 https://www.unglobalcompact.org/news/3381-04-22-2016

175 https://www.cdp.net/en/reports/downloads/1132

176 http://unfccc.int/secretariat/momentum_for_change/items/9261.php

177 http://www.aboutmcdonalds.com/mcd/sustainability/signature_programs/best_practices/best-of-green/innovative-power-purchase-agreements.html

178 http://www.businessrenewables.org/corporate-transactions/

179 https://www.salesforce.com/blog/2015/12/salesforce-renewable-energy.html

180 http://www.repower.net.au/repower-one.html

181 http://farmingthesun.net/lismore/

182 http://www.cleanenergyfinancecorp.com.au/

183 http://arena.gov.au/

184 https://www.google.com.au/green/energy/

185 http://media.virbcdn.com/files/f9/d6e716c56a9b3312-RE100AnnualReport2016_v17.pdf

186 Ibid.

187 Ibid.

188 http://www.lismore.nsw.gov.au/cp_themes/default/page.asp?p=DOCQMM-54-48-20

189 http://media.virbcdn.com/files/f9/d6e716c56a9b3312-RE100AnnualReport2016_v17.pdf
190 http://www.brauunion.at/
191 https://en.wikipedia.org/wiki/Volkswagen_emissions_scandal
192 https://cleantechnica.com/2016/08/18/apple-moves-one-step-closer-100-renewable-energy-worldwide/
193 https://www.lora-alliance.org/
194 http://evo-world.org/
195 https://www.ftc.gov/policy/federal-register-notices/guides-use-environmental-marketing-claims-green-guides
196 https://www.legislation.gov.au/Details/C2015C00019
197 https://www.accc.gov.au/system/files/Your%20consumer%20rights%20environmental%20claims.pdf
198 http://www.v-c-s.org/project/vcs-program/
199 http://www.goldstandard.org/
200 http://www.apple.com/au/environment/climate-change/
201 https://www.google.com.au/green/energy/
202 http://www.greenpeace.org/international/en/Cool-IT-Leaderboard/
203 http://www.lismore.nsw.gov.au/cp_themes/news/page.asp?p=DOC-VXA-06-81-62